17.50

continued on back

Applications of Statistics
to Industrial Experimentation

APPLICATIONS
OF STATISTICS
TO INDUSTRIAL
EXPERIMENTATION

CUTHBERT DANIEL

JOHN WILEY & SONS

New York · London · Sydney · Toronto

Published by John Wiley & Sons, Inc.
Copyright © 1976 by Cuthbert Daniel

Library of Congress Cataloging in Publication Data:

Daniel, Cuthbert.
  Applications of statistics to industrial experimentation.

  Bibliography: p. 289–292.
  Includes indexes.
  1. Experimental design.
  2. Research, Industrial—Statistical methods. I. Title.

T57.37.D36   607'.2   76–2012
ISBN 0–471–19469–7

To Janet

Preface

An experiment is an intervention in the operation of a working system. It is always done to learn about the effect of the change in conditions. Some conditions must be controlled; some, at least one, must be varied deliberately, not just passively observed. To avoid Chantecler's mistake, the variation should not be regular. (You will remember that Rostand's cock thought it was his crowing that made the sun rise.) All industrial experiments are interventions; unfortunately not all are irregularly timed interventions.

It is impossible to make any very general statistical statements about industrial experiments. No claim is made here for the universal applicability of statistical methods to the planning of such experiments. Rather, we proceed by examples and by modest projections to make some judgments on some sorts of industrial experiments that may gain from statistical experience.

Industrial experiments may be classified in several ways that carry implications for statistical thinking. First, I put J. W. Tukey's distinction between confirmation and exploration experiments, which might well be extended by the small but important classification of fundamental, or creative, or stroke-of-genius experiments. This book deals almost entirely with confirmatory experiments, a little with exploratory ones, and not at all with the last type. Confirmation experiments are nearly always done on a working system and are meant to verify or extend knowledge about the response of the system to varying levels or versions of the conditions of operation. The results found are usually reported as point- or confidence-interval statements, not as significance tests or P-values.

A second way of classifying experiments is based on the distance of their objectives from the market. As we get closer to being ready to go into production (or to making a real change in production operations), it becomes more important to have broadly based conclusions, covering the effects of realistic ranges of inputs, operating conditions, on all properties of the product. The farther we are to the right on the God–Mammon scale, the more useful large-scale multifactor experiments are likely to be.

A third classification involves continuity of factors. If most factors in an experimental situation are continuously variable and are controllable at predetermined levels, the whole range of response surface methodology becomes available. These procedures are only cursorily discussed here, since there are already many excellent expositions in print. When many factors are orderable in their levels, but not measurable, the response surface methods become less useful. When many factors are discrete-leveled and unorderable, one's thinking and one's designs necessarily change to accommodate these facts. It is with these latter types of situations that this work is mainly concerned.

A fourth classification distinguishes between experimental situations in which data are produced sequentially and those in which many results are produced simultaneously, perhaps after a lapse of time. Pilot plants, full-scale factory operations, and even bench work on prototype equipment usually produce one result at a time. Storage tests, and clinical trials on slowly maturing diseases are examples of situations that are intrinsically many at a time, not one at a time. They are always multiple simultaneous trials since a long time may be needed to fill in omissions. A very large number of such experiments have been carried out, and dozens have been published. They are strongly isomorphic with the corresponding agricultural factorial experiments. At the one-at-a-time end of this scale I believe but cannot prove that some statistical contribution is to be expected. No examples of completed sets can be given.

Experiments vary in their sensitivity. In some situations the effect of interest Δ is four or more times the error standard deviation σ of the system, so that $\Delta/\sigma \geqslant 4$. In such cases, small numbers of trials (runs, tests, sub-experiments) are required, and replication is supererogatory. This happens most commonly in physical sciences, and in bench work when the experimental setup is familiar and stable. At the other extreme are situations in which $\Delta/\sigma \leqslant 1$, as is common in the biological sciences, including clinical trials, and in work on large-scale, even plant-wide, experiments, where uncontrollable variation is always present and small improvements are commercially important. Statistical methods can be well adjusted to this whole gamut, and the details of this coverage will be given in several chapters.

The book should be of use to experimenters who have some knowledge of elementary statistics and to statisticians who want simple explanations, detailed examples, and a documentation of the variety of outcomes that may be encountered.

CUTHBERT DANIEL

Rhinebeck, New York
March 1976

Acknowledgments

Parts of this work have been discussed and criticized in statistical colloquia at the universities of Princeton, Yale, and Connecticut at Storrs and at the Courant Institute of New York University. Harry Smith and J. S. Hunter have given me valuable detailed criticism of an earlier draft. Also, F. Anscombe, W. G. Cochran, O. Kempthorne, B. H. Margolin, R. W. Kennard, F. Mosteller, H. Scheffé, and H. F. Smith have made searching comments. I am grateful for all of this help and have tried to incorporate it into a better book.

My clients over the past 30 years have provided the ground, the basis, the occasion for all that I may have learned in applied statistics. Without exception they have been patient, intelligent, and generous. I think especially of the research and development staffs of Kellex–Union Carbide Chemicals Corporation at Oak Ridge, of Procter and Gamble, United States Steel Corporation, M. W. Kellogg, General Foods, American Oil Company–Panamerican Petroleum, Itek, Okonite, Interchemical Corporation, Consumers Union, and Technicon Instruments Corporation.

I am deeply indebted also to Ms. B. Shube of Wiley-Interscience and to many others on the staff of that company for their patience and care in editing and seeing this work through the press. It is a pleasure to thank the printers and compositors who have so often seen how to make a clear table from an unclear manuscript. In addition, I want to thank the publisher's reviewers, who gave me many good suggestions.

S. Addelman and the editors of *Technometrics* have given me permission to reprint several tables and one full article (all identified in the text).

Finally, H. Scheffé has provided, over the past twenty-five years, aid and encouragement beyond praise and almost beyond tally. I find that I have four thick folders in my files, all labeled "Scheffé, Current."

C. D.

Contents

Applications of Statistics
to Industrial Experimentation

CHAPTER 1

Introduction

1.1. THE RANGE OF INDUSTRIAL RESEARCH

The connections between scientific research and industrial research are sometimes very close. In studying a new industrial use for the water-gas shift reaction, for example, industrial research workers would depend heavily on the theoretical and experimental results in the technical literature. In producing a new modification of a familiar dyestuff with somewhat improved lightfastness, one industrial organic chemist would start with a careful theoretical study and search for the relevant literature. Another equally able chemist might prefer a wider search of alternatives directly in the laboratory. In attempting to find acceptable operating conditions to make a new petrochemical, it might well be discovered that *no* basis for a theory exists until a considerable volume of laboratory work has been completed.

A wide spectrum of degrees of empiricism already exists, then, in industrial research. The word *theory* is used with entirely different references in different parts of this spectrum. The word may be almost a term of derogation when used by a chemist working on a problem requiring a high degree of empiricism, to describe the work of another who has a good mathematical background but a less sound laboratory foreground. In such contexts the term *in theory, yes* is usually understood to be followed by the phrase *in practice, no.* Contrariwise, the experienced kineticist (even more so, the fresh graduate) may believe that the bench worker should use the term *conjecture*

1

or the expression *set of vague and prejudiced hunches* rather than the fine word *theory* to describe the set of beliefs under which the latter is laboring.

The effects of *some* factors on *one* property of an industrial product may well be broadly guessed or even precisely predicted from available theory. But no industrial product has only one property of interest. It must be stable and inexpensive and small and inodorous and easy to use, and so on, through a list of perhaps 20 attributes. For many of these, little or no theory will be available. Even when theoretical methods might yield correct answers, it may be that no one is available who can use these methods expeditiously. Time will often be saved by simply "getting the data."

Most of my own experience with industrial experimentation has been near the empirical end of the spectrum just indicated, and this bias will show repeatedly in later chapters. The two-level multifactor fractional replicates—and other incomplete two-level factorials—which are one of the principal subjects of this work are quite surely of wide application when a broad range of experience must be accumulated economically in the absence of easily applied theory. Little, but still something, will be said about the prospects for other, more theoretically developed branches of industrial research.

Real differences of opinion on how best to proceed may become very important. Theoreticians may judge that a problem should first be studied "on paper"; laboratory workers may feel certain that the primary need is for more data. Compromises should be considered. Perhaps both views can be implemented at the same time. If the theoreticians can tell the laboratory workers what data they would most like to have, the information may be produced more quickly than either group thought possible. This is so because more can be found out *per run made* or per compound synthesized or per product modification carried out than most experimenters realize.

1.2. SCIENTIFIC METHODS

The research worker is often able to see the results of one run or trial before making another. He may guess that he can improve his yield, say, by a slight increase in temperature, by a considerable increase in pressure, by using a little more emulsifier, *and* adding a little more catalyst. He will act on all four guesses at once in his next run. And yet, in conversation, especially in general or philosophical conversation, he may state his belief in the value of varying *one factor at a time*. Indeed many experimenters identify the one-factor-at-a-time approach as "the" scientific method of experimentation.

Two different phases of research are being confused here. In the very early stages of any problem, *operability* experiments must be done, to see whether *any* yield or other desired property is attainable. After some set of operable or promising conditions has been established, the experimenter is very likely to continue trying simultaneous variation of all factors he thinks may help.

When this no longer works, he may well decide that he must settle down and vary the experimental conditions one at a time, in sequences that are natural for the system in question. This will often involve series of small increments for each of the continuously variable factors. Confusion appears when methods that seem appropriate for the later stage are claimed as valid for the earlier one.

As the process or product gets closer to the market, more and more *conditions and tolerances* turn up as requirements. Toxicity, inflammability, shelf life, and compatibility with dozens of other materials may have to be studied. The tolerance of the product to a wide variety of circumstance of use begins to assume major importance. The research or development technician must now investigate a whole set of new conditions. He must be able to assure the producing and marketing divisions of his company that the product can be guaranteed safe, efficient, and operable under a range of conditions not studied when it was first being considered and developed.

Because of the shortage of available technicians, because of the entire lack of any theory for some properties, because of the multiplicity of factors that *may* influence a product, and because of the other multiplicity of factors to which it must be insensitive, industrial research often differs widely from pure or basic research. In particular, more factors must be studied, and so it is often said, and rightly, that more data must be taken in industrial research problems than in pure research ones.

1.3.   MAKING EACH PIECE OF DATA WORK TWICE*

It does not follow that the enormous amounts of data often accumulated in industrial research laboratories are entirely justified. Most experimenters, and most research directors too, I believe, have assumed that each piece of data can be expected to give information on the effect of one factor *at most*. This entirely erroneous notion is so widespread and so little questioned that its correction should start right here with the simplest possible example to the contrary.

A chemist has two small objects to weigh. He has a double-pan scale of fair precision and of negligible bias and a set of weights with excellent calibration. He would like to know the weight of each object with the best precision possible. He is to make *two* weighings only. His experience, habits, and common sense conspire to tell him to weigh one object (call it *P*) and then to weigh the other, *Q*—carefully of course. For each object there will be one weighing, one piece of data, one weight.

There is, however, a way to find the weight of each object as precisely as if *it* had been weighed twice and the two weighings averaged. To do this each

* This expression is due to W. J. Youden.

object must indeed be weighed twice. But we are allowed only two weighings in all. Hence each object must be on the scale pans twice. If the two objects are put in one pan and weighed together, we get an estimate of the sum of the two weights. To separate the components we must either weigh just one, or else find their difference. By placing one object in one pan and one in the other, we can, by balancing with the calibrated weights, find the difference. Calling the sum of the weights $S = P + Q$, and the difference $D = P - Q$, we see that the average of S and D measures the weight of P only, since Q is exactly balanced out. Similarly, the average of S and $-D$ measures the weight of Q with P exactly balanced out. We have then weighed each object twice, in two weighings, each with the precision of two averaged weighings.

The disadvantage of this "weighing design" is that no information is available until all the data are in. The reward for the delay is, in this case, the double precision. The moral, to be given extended emphasis and development later, is that each observation can be made to yield information on two (or more) parameters. Indeed the number of times that each observation can be used increases steadily with the number of observations in each balanced set. What is required is *planning*. In most cases, little or no information is extractable along the way. Finally a computation, usually quite simple, must be made to extract all the information at once.

The pronounced improvement of the (S, D) pair of weighings over the (P, Q) set becomes a minor matter when compared with the gains that are attainable when larger sets of weights or any other measurements are to be estimated. The simplest case was used here as an example that does not appear to have been mentioned since it was first pointed out by Hotelling [1942].

1.4. FIRST STAGES IN PLANNING INDUSTRIAL EXPERIMENTS

The stated aims of an industrial experiment are not the same at all of its stages, but the same broad desiderata seem to emerge repeatedly. We always want to know whether an effect holds fairly generally, and whether an apparent lack of effect of some factor is a general lack. Fisher's determined emphasis on the importance of a broad base for scientific inferences can never be forgotten. It is not a counsel of perfection but rather a sine qua non for good industrial research.

Some experimenters believe that they must be able to judge *early* which factors are going to be influential. They foresee, or think they do, that the experimental program will become unmanageably large if all factors are admitted for detailed study. But if factors are dropped from the active list too early, on the basis of relatively small numbers of data, it may take the research worker a long time to get back on the right track.

It is better to write down quite early a full list of all the factors that might influence the desired properties of the product under development. A valuable exercise in planning an attack on a new problem is to prepare a cross tabulation of all potential factors by all interesting properties of the product or process. This "influence matrix" should be a dated record of the experimenter's opinions about the effect(s) of each independently controllable variable on each property. Its use is discussed in Chapter 9.

Within the limits of practicability it is desirable to look at each factor's effects under a wide range of conditions or levels of the other factors. A stable effect, even at zero, over a wide range of settings of the other factors is reassuring because broadly based. On the other hand, if the effect of some factor varies, perhaps even changes sign depending on the settings of the others, this information is important and should be known early. Balanced or nearly balanced sets of runs provide the easiest way to learn about these situations.

Perhaps the major departure of this work from others with similar subject is its attitude toward the assumptions that are usually made before experimentation is started. The standard assumptions of most statistical treatments are as follows:

1. The observations must be a fair (representative, random) sample of the population about which inferences are desired.
2. The observations are of constant variance (or at least the variance must be a known function of the independent variables), are statistically independent, and are normally distributed.
3. Few or no bad values will be produced, and few missing values.

Assumption 1 is for the experimenters to guarantee. The three parts of assumption 2 can often be verified, or at least refuted, by the data themselves. Responding to the myriad ways in which data fail to meet these requirements will be a major part of the effort. Assumption 3 is violated in a large number, perhaps 30%, of all industrial experiments. Methods are given for spotting bad values, and for drawing valid conclusions, though often with reduced precision, in spite of these defects.

1.5. STATISTICAL BACKGROUND REQUIRED

I assume that the research worker reading this book knows a few of the fundamentals of applied statistics. Foremost among these are the following:

1. The prime requirement for drawing any valid inference from experimental data is that the inferrer know something about the way in which

the data (a sample) represent nature (the population). The prime requirement for the validity of any conclusions drawn from the study of experimental data is that the data be a real sample of the situations to which
the conclusions are to apply.

2. The basic terms—*statistic, parameter, sample mean, sample standard
    deviation, standard error of a mean, regression coefficient, least-squares
    estimate*—should all be familiar. They will all be defined and described,
    but if the reader is encountering many of them for the first time, he will
    not find these pages easy reading.

3. The most pervasive generalization in the whole of statistics is the Central
    Limit Theorem. Its effect is to make averages of independent observations
    more nearly Gaussian in their distribution than the error distributions
    of the single observations. Since a large proportion of the parameter
    estimates we make are averages, the central limit theorem must be
    working for us a large part of the time. This comforting circumstance
    cannot account for the apparent "normality" we will repeatedly find in
    residuals, however, since they are heavily dependent on the single ob-
    servations themselves. For these we must believe that a considerable
    number of small additive, nearly independent factors are responsible.
    No quantitative knowledge or application of the theorem is ever neces-
    sary. It simply operates, like a law of nature, but, unlike other laws,
    generally in our favor. The reader is referred to Cramér [1946] for
    an illuminating discussion of the central limit theorem and of its ante-
    cedents.

1.6. DOING THE ARITHMETIC

Many research engineers and industrial scientists are repelled by the mo-
notonous and extensive arithmetic that statistical texts and handbooks seem
to demand. My sympathies are with them; much of this drudgery is unnecessary. Nearly all the arithmetic in this book has been done by hand,
perhaps on a desk calculator. Intelligent coding and rounding are of the
essence and frequently result in reducing time, as well as errors, to a small
fraction of their former magnitudes.

When 10 or more experiments (or responses in a single experiment of size
16 or larger) must be analyzed, time will be saved if the standard algorithms
(for the analysis of variance, for Yates's method in 2^n plans, for partially
balanced incomplete blocks) are available on a computer. *Do not consider
any program that does not compute and print residuals automatically, preferably rounded to two digits.*

The plotting of cumulative empirical distributions of residuals on a
"normal" grid is again a tedious job when done as proposed in the few

textbooks that mention it. But when the number of points is large, the job can be greatly shortened. Standard grids for $N = 16, 32$ are given that require no calculation of probability points. All large computers have been programmed to a fare-thee-well to make approximate plots without special peripheral equipment, and only approximate plots are needed. I have found too that, when the number of points exceeds 100, it is usually necessary to plot only the largest 25 or so (including both ends). As soon as the plotted set "point" straight through the 50% point, there is no need to continue plotting.

1.7. SEQUENCES OF EXPERIMENTS

The *analysis* of sequences of agricultural experiments has been studied extensively by Yates and Cochran [1957, pages 565 ff.], and much can be learned from this work. The *design* of sequences of industrial experiments is much less fully developed, although economical augmentation of early experiments seems to be crucial in industrial research. The earliest work in this area was by Davies and Hay [1950]. Less clear, but more economical, augmentations were published in 1962 [Daniel]. [Although trend-robust plans (Chapter 15) are carried out in sequence, they are not really adaptive designs but have to be carried all the way before effects can be estimated.]

1.8. THE FUTURE OF THE DESIGN OF INDUSTRIAL EXPERIMENTS

Major new developments in the design of industrial experiments seem to me to await the appearance of well-educated statisticians who want to work in close touch with industrial scientists. Many mathematical statisticians are under the illusion that they and their graduate students are writing for a future which they forsee without benefit of detailed knowledge of the present. A tiny proportion of their work may be remembered 20 years from now.

As in the past, many developments will come from scientists and engineers with extensive experience in industrial research. But we need in addition a cohort of modest graduate statisticians who recognize the productiveness of going directly to industrial scientists to find out just how they do their research. Far too many graduates, and even some senior statisticians, are willing if not anxious to tell scientists how to plan their experiments, in advance of knowing just how such work is now done. There are, fortunately, a few outstanding exceptions. I think especially of the work of Box, Lucas, Behnken, W. G. Hunter, N. R. Draper, and their associates on "nonlinear" design. A shortcoming of this book is its lack of any treatment of these plans—an omission due to my own lack of experience with them.

CHAPTER 2

Simple Comparison Experiments

2.1.   AN EXAMPLE

The example is taken from Davies [1956, Ed. 2, printing 1971, pages 12–18]. Only one criticism is to be made, and that with some hesitation, since this is the fundamental work on industrial experimentation (from which I for one have learned more than from any other book).

We quote from Davies, Section 2.21:

> The experiment was required to test whether or not treatment with a certain chlorinating agent increased the abrasion resistance of a particular type of rubber. The experimenter took ten test-pieces of the material and divided each piece into two. One half was treated and the other half was left untreated, the choice of which half of the specimen should receive the treatment being made by tossing a coin. The abrasion resistances of the ten pairs of specimens were then tested by a machine, the specimens being taken in random order.

Perhaps most experimenters would prefer to call such a collection of data a *test*, so as not to invoke the grander connotations of the term *scientific experiment*. It is not clear from the description or from later discussion (page 43, Figure 2.5) whether all 10 specimens were taken from one sheet of rubber. Since we need a straw man for this discussion, let us assume that the 10 were indeed a random sample from a single sheet. Randomization of the choice of half piece for chlorination plus random allocation of sample points in the sheet of rubber have guaranteed that any difference found and judged to be real has a good chance of being confirmed if measured over the whole sheet.

But the data come from one sheet of rubber. The pains taken to obtain precise and "unbiased" data have resulted in our getting into our sample only a small corner of the population about which we suppose the experimenters wished to make a statement. If other sheets, from other batches or from other manufacturers, would respond differently to this treatment, then we will be misled by the results from our single sheet. The conclusion from these data, expressed as a "95% confidence interval on the true effect of chlorination," is applicable only to the average effect for all parts of the sheet sampled.

It may well have been that only one sheet of the particular type of rubber under study was available. But if more than one could have been sampled, more valid conclusions would have been reached by sampling them all. Sampling 10 sheets chosen at random would be best. Moreover, if several manufacturers produced this type of rubber, still greater validity could have been guaranteed by sampling all, even if only with one specimen from each.

An important function of the design statistician is to give the experimenter pause, before he takes his data, in order to help him avoid the commonest of all mistakes in experimental (and testing) work. *This is the mistake of premature generalization.* It is most frequently caused by assuming that the data are a proper sample of a wider population than was in fact sampled.

It is time for Sermon I: "The Contribution of the Statistician (S) to the Experimenter (E)." It will be a short one.

The major contribution of S to E is to help him obtain more valid, that is to say, more general, more broadly based, results. It will often happen that, when this point has been adequately covered, no need is felt, or time available, for repeated measurements under closely matched conditions. The most useful replication will be that which best samples the population of conditions about which E wants to make inferences. In this sense, the best replication is done under different conditions, not under the same conditions.

Although we are by no means through with simple comparisons, the experimenter will perhaps see a new answer to the question usually asked rhetorically: "What can statistics do for me?" The statistician reader, in turn, may give a somewhat new emphasis in answering the question, "What is the key assumption underlying the valid use of a confidence-interval procedure?" His response should be that we have a set of observations with known correlations, preferably zero, *sampling the population whose parameters we want to estimate.*

2.2. *THE* EFFECT OF A FACTOR?

The patient experimenter has been thinking that this is all very simple minded and not really what he had hoped for. "Real problems are more complicated. Even chlorination is not that simple. There are many chlorinating agents, there are degrees of chlorination, and so on. Surely we are not to be told that *the* effect of chlorination is going to be decided on so flimsy a basis, with all that fuss about sampling rubber and nothing about sampling the conditions of chlorination."

Just as we tried to broaden the base of our experience by better sampling of the particular type of rubber, so too must we sample better, and systematically, the conditions for chlorination. If we do not, we may report "the effect" because of our choice of conditions, even though other levels, not greatly different, might show larger effects. The rest of this book is concerned with various aspects of this question.

CHAPTER 3

Two Factors, Each at Two Levels

3.1. INTRODUCTION

The study of the effects of varying a single factor is usually only a prelude to the study of the effects of varying several factors. Only minimal generality is gained by repeated variation of a single factor, with everything else held constant. This practice is commonly justified by the claim that "we have to start somewhere." We do indeed.

Researchers faced with serious scientific problems have long records of success in choosing the most important factor and then studying it thoroughly. Later, if not sooner, however, they usually need to learn about the simultaneous impact of two or more independent variables, or at least about the response to variation of one factor under more than one set of conditions. If one of two catalysts proves definitely better than the other in cracking a particular petroleum feedstock, it is inevitable that the experimenters will want to know whether it is also better, and, if so, to the same degree, in cracking another stock. They then have before them the simplest two-factor plan, a 2^2.

13

The distinction between factors with qualitatively different levels (or versions) like the two just mentioned and "quantitative" factors (like pressure, temperature, chemical concentration) will be made frequently, but at the moment is not important.

3.2. FACTORIAL REPRESENTATIONS

Suppose, to fix fundamental ideas, that each of two factors, A and B, has been varied, and that the results (responses, y's) are as follows:

		B	
		0	1
A	0	66	82
	1	44	60

or, in general,

		B	
		0	1
A	0	y_{11}	y_{12}
	1	y_{21}	y_{22}

We hardly need to know the error standard deviation for the "data" on the left to be able to judge that we have before us the ideal case. Varying A from its low to its high level has produced the same change in response ($44 - 66 = 60 - 82 = -22$) *at both levels of B*. Varying B from its low to its high level has produced a change in response of $+16$ *at both levels of A*. We can speak, then, of the *additivity* of effects of factors A and B and can safely symbolize the situation by writing, first generally,

$$(3.1) \qquad Y_{ij} = b_0 + b_1 x_{1i} + b_2 x_{2j},$$

where $x_{11} = 0$, signaling low A;
$\quad x_{12} = 1$, signaling high A;
$\quad x_{21} = 0$, signaling low B;
$\quad x_{22} = 1$, signaling high B;
$\quad i$ indexes the levels of A, i.e., $i = 1, 2$;
$\quad j$ indexes the levels of B, i.e., $j = 1, 2$;
$\quad b_0$ is the value of Y_{11}, i.e., at the low levels of both A and B;
$\quad b_1$ is the increment in Y caused by changing the level of A: it is the "effect of varying A";
$\quad b_2$ is the increment in Y caused by changing the level of B: it is the "effect of varying B."
For the data given above, this becomes

$$Y_{ij} = 66 - 22x_{1i} + 16x_{2j}.$$

We will consistently use Y (capital) to indicate fitted or predicted or "regression" values, and y (lowercase) to indicate single observed results. Variables such as x_1 and x_2 in Equation 3.1 that can take only two values (here 0 and 1) are called *indicator* (or *dummy*) variables.

In an equivalent nomenclature, used because of its greater symmetry, we may put $x_1 = -1$ at low A, and $= +1$ at high A, and similarly for x_2. The effects are now measured as deflections up and down from the general average $(y_{..})$ of the results. The *half* effects are commonly symbolized by their corresponding letters, with carets superimposed to indicate that these are estimates, not parameters. The fitting equation is now written as

$$(3.2) \qquad Y_{ij} = y_{..} + \hat{A}x_{1i} + \hat{B}x_{2j}$$

or, for the imaginary data just given, as

$$(3.3) \qquad Y_{ij} = 63 - 11x_{1i} + 8x_{2j}.$$

All three constants in (3.3) are calculated from the data:

$$y_{..} = \tfrac{1}{4}[y_{11} + y_{21} + y_{12} + y_{22}] = \tfrac{1}{4}[66 + 44 + 82 + 60] = 63$$
$$\hat{A} = \tfrac{1}{4}[(y_{22} + y_{21}) - (y_{12} + y_{11})]^*$$
$$= \tfrac{1}{4}[(60 + 40) - (82 + 66)] = \tfrac{1}{4}(104 - 148) = -\tfrac{44}{4}$$
$$= -11,$$
$$\hat{B} = \tfrac{1}{4}[(y_{12} + y_{22}) - (y_{11} + y_{21})]^*$$
$$= \tfrac{1}{4}[(82 + 60) - (66 + 44)] = \tfrac{1}{4}(142 - 110) = \tfrac{32}{4}$$
$$= 8$$

In tabular form, which avoids some repetition, we have

		B			
		0	1	Row Average	$\hat{A}$-Effect
A	0	66	82	74	+11
	1	44	60	52	−11
Column Average		55	71	63	
$\hat{B}$-Effect		−8†	+8		

† $\hat{B} = 55 - 63 = -8$.

Statisticians are wont to write (3.3) in a third way, which is more useful when there are more than two levels of each factor:

$$(3.4) \qquad Y_{ij} = y_{..} + \hat{\alpha}_i + \hat{\beta}_j,$$

where now the i levels of factor A can be numbered $1, 2, \ldots,$ and so also for j.

---

* These expressions are our first examples of contrasts, that is, of linear functions of variables—here y_1, y_2, y_3, y_4—whose coefficients sum to zero. Thus $1 + 1 - 1 - 1 = 0$. All effects are estimated by contrasts.

Accelerating our snail's pace, let us suppose that a 2^2 has been done but that the results are as follows:

		B 0	B 1	$\hat{\alpha}_i$
A	0	67	81	$+11$
	1	43	61	-11
$\hat{\beta}_j$		-8	$+8$	

Note that the $y_{..}$, $\hat{A}$, and $\hat{B}$ are the same for these data as for the earlier set. But now there is a discrepancy of ± 1 in each cell. We can represent the decomposition of the observations into four parts as follows:

$$(3.5) \quad \begin{bmatrix} 67 & 81 \\ 43 & 61 \end{bmatrix} = \begin{bmatrix} 63 & 63 \\ 63 & 63 \end{bmatrix} - 11 \begin{bmatrix} -1 & -1 \\ +1 & +1 \end{bmatrix} + 8 \begin{bmatrix} -1 & +1 \\ -1 & +1 \end{bmatrix} + 1 \begin{bmatrix} +1 & -1 \\ -1 & +1 \end{bmatrix},$$

$$(3.6) \quad y_{ij} = y_{..} + \hat{A}\, x_{1i} + \hat{B}\, x_{2j} + \widehat{AB}\, x_{1i}x_{2j}.$$

Thus the effects $\hat{A}$ and $\hat{B}$ do not give an exact representation of the four observed y-values. If $\sigma(y)$ is of the order of 1, this discrepancy is put down to error.

If $\sigma(y)$ is 0.1, we are confronted by real lack of fit. The factors A and B are said to interact—they no longer operate exactly additively. We can add a fourth term to our equation, which will obviously have to be of second order, and equally obviously cannot be of the form x_1^2 or x_2^2 (these are meaningless) and so, again obviously, must be of the form Cx_1x_2. We will make the new constant easier to remember by writing it as $\widehat{AB}$, and so the new term is $\widehat{AB}x_{1i}x_{2j}$. It is called a two factor interaction. In the example we are flogging, $\widehat{AB} = 1$. This can be calculated directly from the data by using the tiny matrix

$$\begin{bmatrix} 1 & -1 \\ -1 & 1 \end{bmatrix}$$

at the end of (3.5). Thus we have

$$\widehat{AB} = \tfrac{1}{4}(y_{11} - y_{21} - y_{12} + y_{22}) = \tfrac{1}{4}[(y_{22} - y_{12}) - (y_{21} - y_{11})],$$

where the second form makes it clear that $\widehat{AB}$ depends on the difference between the A-effect at high B and the A-effect at low B. The term "two-factor interaction" will be abbreviated as 2fi throughout this book.

In the representation of (3.4) we can now write

$$(3.7) \quad Y_{ij} = y_{..} + \hat{\alpha}_i + \hat{\beta}_j + \hat{\gamma}_{ij},$$

where the first three terms are unchanged in meanings and values from (3.4).

The $\hat{\gamma}_{ij}$ nomenclature is widely used [Brownlee 1965, Scheffé 1958] but is only minimally useful for the 2^2 since it contains only one degree of freedom (d.f.), although there are four i, j combinations.

Adding two more self-explanatory modes of specification of the four "experimental settings" for the 2^2, we have Table 3.1.

TABLE 3.1.

FOUR ALTERNATIVE REPRESENTATIONS OF THE EXPERIMENTAL CONDITIONS
FOR THE 2×2 FACTORIAL PLAN, THE 2^{2+0}

Run No.	Uppercase	Digital A B	Symmetrical Coordinates x_1 x_2	Lowercase	Response
1	A_0B_0	0 0	-1 -1	(1)	y_1
2	A_1B_0	1 0	$+1$ -1	a	y_2
3	A_0B_1	0 1	-1 $+1$	b	y_3
4	A_1B_1	1 1	$+1$ $+1$	ab	y_4

The four nomenclatures are designated here as uppercase, digital, symmetrical, and lowercase. They are, of course, ways of specifying the experimental conditions, not the responses. The uppercase symbols give levels of factors by subscripts; the digital symbols are two-digit numbers: 00, 01, etc., and are simply the subscripts of the uppercase symbols; the symmetrical coordinates are the levels of x_1 and x_2 restricted to be $+1$ or -1. The lowercase symbols are the most compact, since they use the absence of a letter to indicate the lower level, and the presence of a letter to indicate the higher level, of the corresponding factor.

Returning to our "concrete" example, the reader who is following closely will see that the fitting equation for these four values in the factorial representation is

(3.8) $$Y_{ij} = 63 - 11x_{1i} + 8x_{2j} + 1.0x_{1i}x_{2j}.$$

3.3.   YATES'S ALGORITHM FOR EFFECTS IN THE 2^2

We start with the ridiculously simple case of the 2^1, that is to say, a *one-factor two-level experiment*, and consider Table 3.2. The experimental conditions are indicated by (1) and a, and the results by y_1 and y_2. The columns headed T and (A) show how the data would be treated to find twice the mean and twice the A-effect. To make a short story long, we add the pair and then subtract y_1 from y_2. To get the two equation constants, we divide

TABLE 3.2.
TABLE OF SIGNS AND YATES'S ALGORITHM FOR THE 2^1

Spec.	Obs.	$T = 2y$	$(A) = 2\hat{A}$	Obs.	Column 1	Name
(1)	y_1	$+1$	-1	y_1	$y_2 + y_1$	T
a	y_2	$+1$	$+1$	y_2	$y_2 - y_1$	(A)

by 2^1. The fitting equation is then

$$Y = y_{.} + \hat{A}x_1,$$

where x_1 is the indicator variable that takes the values -1 for low A and $+1$ for high A. We have omitted the fussy subscript i here.

For the 2^2 we also handle the numbers one pair at a time. As Table 3.3 shows, adding numbers in pairs removes the effect of factor A. These sums are entered in the first two lines of column 1. We recover the simple A-effects by taking differences between the pairs as in the last two lines in column 1. From the name of each of the four entries in column 1, we see how these must be combined to give our effects (each multiplied by 2^2). Adding the pairs of column 1, we get T (the total) and (A), the *contrast-sum* that measures $4\hat{A}$. These are shown in the first two lines of column 2 of the next table, Table 3.4. We see, too, that the difference between the low-B sum and the high-B sum will give the total B-effect $(= 4\hat{B})$, and that the difference between the two A-effects will give $4\hat{A}B$.

TABLE 3.3.
PARTIAL COMPUTATION OF FACTORIAL EFFECTS
FOR THE 2^2 BY YATES'S ALGORITHM

Spec.	Column 1	Name of Sum or Difference
(1)	$a + (1)$	Low-B sum
a	$ab + b$	High-B sum
b	$a - (1)$	A-effect at low $B = \hat{A}_1$
ab	$ab - b$	A-effect at high $B = \hat{A}_2$

Written out in such detail, these directions may not seem timeworthy, but this algorithm is wonderfully compact when $n = 3$ or more, and we will show further uses for it later. In Table 3.5 the data that follow (3.4) are put through this computation for practice. The panel on the right is an exercise for the reader. [In this table, as in many others presented subsequently, (0), (1), (2), etc., in the heading designate column 0, column 1, column 2, etc.]

TABLE 3.4.

COMPLETE COMPUTATION OF FACTORIAL EFFECTS FOR
THE 2^2 BY YATES'S ALGORITHM

Spec.	Column 1	Column 2	Name
(1)	$a + (1)$	$ab + b + a + (1)$	$T = 4y_{..}$
a	$ab + b$	$ab - b + a - (1)$	$(A) = 4\hat{A}*$
b	$a - (1)$	$ab + b - a - (1)$	$(B) = 4\hat{B}$
ab	$ab - b$	$ab - b - a + (1)$	$(AB) = 4AB$

* The symbol (A) will be used in this work to denote the
total effect of A. As mentioned in the text, it is called a
"contrast-sum" and is always $2^n \hat{A}$.

TABLE 3.5.
SAMPLE COMPUTATION FOR A 2^2

Spec.	(0) Obs.	(1)	(2)	(3) (2) ÷ 4	(4) Name	Obs. (An Exercise)
(1)	67	110	252	63	$y_{..}$	11
a	43	142	-44	-11	$\hat{A}$	11
b	81	-24	32	$+8$	$\hat{B}$	35
ab	61	-20	4	$+1$	AB	51

We can see how well the equation without the interaction term fits the
data by reversing the algorithm and setting $AB = 0$. This is done most
simply by writing the effects in *inverse* order, carrying through the same set of
additions and subtractions, and reading off the fitted values in *inverse*
standard order, as in Table 3.6.

TABLE 3.6.
REVERSAL OF YATES'S ALGORITHM TO COMPUTE FITTED VALUES
FROM EFFECTS

Effect	(0)	(1)	(2) = Y	Spec.	y	$d_y = y - Y$
$\hat{A}B$	0	8	60	$\hat{ab}$	61	1
$\hat{B}$	8	52	82	$\hat{b}$	81	-1
$\hat{A}$	-11	8	44	$\hat{a}$	43	-1
$y_{..}$	63	74	66	$(\hat{1})$	67	1

The residuals, $d_y = y - Y$, are forced to be of the same magnitude and to have the signs of (AB) because we are fitting three constants to four observations. We know that the addition of an interaction term to our fitting equation will give us an exact fit. There is then but one d.f. in the four residuals, rather than four separate measures of lack of fit to the additive model.

3.4. INTERPRETATION OF A FACTORIAL EXPERIMENT WHEN INTERACTIONS ARE PRESENT

As we glance at the factorial representation of our 2^2 given by (3.8), we may think that the last term, having a notably smaller coefficient than either of its predecessors, should be dropped. Expressing the same idea in another, equally tendentious way, we may say that the equation using only the additive main effects represents all data points with residuals of ± 1. And yet, and yet, we must not forget that the two simple A-effects $\hat{A}_1$ and $\hat{A}_2$, -24 and -20, may for some situations be seriously different. If the average random error of observations is 0.1, and if the experimenter has a physical model that requires additivity of the effects of A over the range of B under study, then the data have sufficed to reject his model.

Our example is too small, as well as too fictitious, to interest us further for its own sake. But a warning and an aid to clarity are in order. Many who use the effects-and-interactions mode of description do not notice that an apparently small interaction may have serious consequences.

Representing the ratio of the two simple effects, $\hat{A}_1/\hat{A}_2$, by r, and the ratio of the interaction to the average main effect, $\hat{AB}/\hat{A}$, by f, we have

$$r = \frac{1 - f}{1 + f}, \quad \text{or, if } f = \frac{p}{q}, \text{ then} \quad r = \frac{q - p}{q + p}.$$

(This formula has the pleasant property that it remains true even if one forgets which ratio is r and which f.) We see that, if $f = \frac{1}{2}$, $r = \frac{1}{3}$. Thus we might well find A significant and AB not, and so might ignore the fact that the data are "trying to tell us," namely, that the A_1-effect is only one third of the A_2-effect.

When the interaction is as large in magnitude as a main effect, we have $f = 1$, and so $r = 0$. In words, A has no effect at one level of B, and all of its effect (and so twice the average, $\hat{A}$) at the other level of B. There are many examples of this in the published literature, but none has been pointed out explicitly.

When the 2fi (between two two-level factors) is of nearly the same magnitude as *both* main effects, there is again a better way to describe the situation.

Suppose that $-A \doteq B \doteq AB = 1$. Then the average responses will look like this:

$$B$$

$$A \begin{array}{|cc|} \hline 1 & 1 \\ -3 & 1 \\ \hline \end{array} \quad \text{with mean zero,}$$

or like this:

$$B$$

$$A \begin{array}{|cc|} \hline 21 & 21 \\ 17 & 21 \\ \hline \end{array} \quad \text{with mean 20.}$$

There is one contrast*, not three, and one sentence that describe the situation completely. The contrast is $[(1) + b + ab - 3a]$; the sentence is: "The condition a is adverse (or advantageous) and the other three combinations of levels of A and B are indistinguishable." This appeared in the classic 2^5 experiment on beans given by Yates [1937]. In his symbols, $-S \doteq K \doteq SK$, and all three were significant. It might well happen in another case that all three effects taken singly were nonsignificant, but that the contrast given above was highly so.

3.5. INTERMEDIATE SUMMARY

The summary takes the form of Sermon II. When you find interactions that approach in magnitude one or more of their component main effects, you should examine the combined impact of the effects *and* the 2fi by the reverse Yates's algorithm. You will sometimes find that all taken together pile up in some part of factor space to produce very large (or very small) values there and little or no difference elsewhere. As a general rule: *If an interaction is one-third or more of a main effect, do not report results in terms of main effects and interactions.* Interactions are only a statistical (and hence descriptive) model's way of telling you that the simple additive model is not working. What *is* working can sometimes be seen by going back to the data or to the values produced by the fitted model. Interactions are then a sort of lack of fit; they are residuals from the fitting of an additive model. They do not always tell us very clearly what to do next in the 2^2 since the interaction appears in all four cells, but we will use them in many more ways in more complex situations.

---

* The reader will remember that a contrast is a linear function, usually of responses, y_i, of the form $\sum_1^n c_i y_i$, where $\sum_1^n c_i = 0$. The c_i are exact constants.

Post Scriptum. There is still another method of reporting interactions (and main effects) that is even more obscure than just the giving of their estimates, and that is to assign a "mean square" to each, to judge their significances individually and to leave the matter at that. This practice will be documented and deplored later.

3.6. THE REPLICATED 2^2

3.6.1. General Remarks on Replication

We have usually assumed that the average random error (standard deviation) of observations is negligible compared to all effects found; otherwise we have assumed that the average error is exactly known. These unrealistic assumptions were made in order to expatiate on the definitions of the factorial parameters, on their interpretations, and on their limitations.

But σ is usually not known and usually must be estimated from data, most simply by replication. The two advantages of a properly replicated 2^2, a 2^{2+r} then, are that a current estimate of the standard deviation σ is obtained, and that each parameter is estimated with smaller variance $\sigma^2/4r$. Further replication will give better and better estimation of error and more and more precise parameter estimation.

Most readers will know what is meant by proper replication. If care has been taken to randomize the allocation of experimental units to "treatments," that is, to the four specified experimental conditions, the replication is quite surely proper.

Randomization is a form of insurance against two sorts of *bias*. An unrandomized experiment may give biased parameter estimates and a biased estimate of the error variance. The experimenter may then carry out entirely meaningless tests of significance, and he may compute "confidence intervals" which are wrongly centered and of incorrect width.

The hazards of nonrandomized tests or experiments are widely different in different sciences, even in different laboratories. It does seem to be the case that a very large part of all scientific and engineering data (perhaps 90%) is taken under nonrandomized conditions. We cannot simply condemn all these experiments and all these data as worthless. Let us rather indicate some of the conditions that make randomization more or less needful.

3.6.2. Limitations of Randomization

Although the "state of the art" is not a quantifiable measure, there can be little doubt that some fields of research exist which are steady enough and precise enough to advance for long periods with no need for randomization. A classic example of *repeated* data that required no randomization whatever is provided by Michelson's nineteenth century measurements of the velocity of light. They were taken in sequences of several hundred over several years, and on several experimental setups. Even though later colleagues were not satisfied, these measurements were "all right" and gave a durable estimate of the mean velocity of light in air, along with an excellent estimate of the random error of the measurements. As Walter Shewhart told me long ago, "the set of several hundred measurements were in as good a state of statistical control" as any set he had seen. By "a state of statistical control" Shewhart meant that successive observations evidently varied only by random independent disturbances of zero average (no drift) and of constant variance.

As a more current, but imaginary, example a physical chemist who wishes to check or to modify Benedict's equation of state for a new gas mixture will surely standardize or calibrate his equipment with familiar gases or mixtures, and will then not look back by making further repetitions. He will vary pressure, temperature, or even trace components in the order that is technically most convenient. If his data permit satisfactory estimation of all free parameters in the equation, with perhaps 10 or 15 extra points to judge goodness of fit and to check for drift, no need for randomization is apparent.

To take an example at the other extreme, in which it is apparent that several biased estimates of parameters were published, the 15 reports on the "solar unit" (average distance of the earth from the sun) appearing between 1895 and 1961 each gave an estimate of this unit together with an estimate of its error, or at least "spread" (see Youden [1962], McGuire et al. [1961]). Each new estimated value lies outside the spread given by its predecessor! It is not easy to see how randomization could have been used to get a fairer estimate of error.

When a detailed physical (or chemical or biological or psychological) mathematical model exists, which diverges clearly from some alternative or competing model, it may well happen that a single run will be decisive. Good experimenters are sometimes able to make tests under just a few differing conditions that provide decisive evidence when the expected difference is large. A large difference is of course one that is a large multiple of its standard error. In 1912 the estimated advance in the perihelion of Mercury that was not accountable for by Newtonian theory was 43 seconds per century, not a large amount by most standards. But this difference was 10 times its estimated standard error, and that was enough to justify two expeditions to the tropics (to make the first tests of the theory of relativity).

When we come to great experimenters—and we do not come to them often, or they to us—the requirement of randomization is derisible. Their reportable experiments are nearly always crucial; results must strike all (well, nearly all) competent readers "between the eyes." One or more who are not convinced are likely to plan and carry out an experiment that will supply striking disproof or, possibly, confirmation. Thus these workers do practice serious and not trivial replication, since an experiment repeated in another laboratory is a more severe test of real effects than is any repetition in a single laboratory. This is hardly done, however, to conform to some statistical canon of unbiasedness. It has been part of the scientific code for centuries that one man's work must be verifiable by another's, at least by some other's.

We now come to experimental situations in which randomization, although theoretically desirable, is not decisive, not needed, perhaps not even sensible. Suppose that a large number of trials, (say 100) have been made with careful randomization, and that 100 more have been made with no such precaution. The sets come from similar but not exactly duplicate experiments. If analysis of both sets shows no appreciable difference in error structure, it would be doctrinaire to insist that all future trials be randomized. Perhaps a fraction, say one quarter, should be, just to keep a rough monitor on the stability of the system under study. This is once more what Shewhart demanded when he required evidence that a system was currently *in a state of statistical control.*

We do not suggest that randomization be ignored just because "things appear to be going along all right." We may not know how nearly all right things have been going until some serious randomized trials have been carried out. We do suggest that randomization, although generally sufficient, may not always be necessary. But this decision requires evidence, not just optimism.

There are many cases in which randomization is difficult, expensive, inconvenient. The random allocation of differing experimental conditions to experimental units is sometimes upsetting to the experimenter. This reaction may be much more than just a natural response to an unfamiliar device. If the system under study takes a long time to come to equilibrium after a sudden willful change in the level of some factor, then experimental work, and even plant production, may be slowed to an unacceptable rate by such a change. Similarly, if a complex system must be partially dismantled to vary some structural factor, it is not likely that random variation of that factor will be permitted. The statistician's way to accommodate such factors is of course to use "split-plot" or "partially hierarchal" designs. These plans, discussed in Chapter 16, have the discouraging property that the effects of the easy-to-vary factors are always more precisely determined than are the effects of the hard-to-vary factors.

Perhaps this paragraph should be entitled "Conditions under Which Nothing Can Be Done." Agricultural experimenters cannot take a random sample of years in which to compare varieties or conditions of cultivation; sometimes they cannot even get a random sample of seeds or test animals. Blast-furnace operators cannot assign different coke charges to a furnace at random. Lung-cancer researchers cannot randomly assign cigarette loadings to human subjects. Examples can be given from many other fields.

There are, to be sure, ways of randomizing stupidly, even disastrously. If one treatment is applied to one field "randomly" and another treatment to another field, the precision of within-field comparisons may be lost. One clinical treatment applied in one research hospital cannot usually be compared safely with another treatment in another hospital.

We now mention two more experimental situations (to be considered later in more detail) in which randomization of the usual sort is contraindicated. *When* the experimenter knows that his random error is small compared to the effects and interactions of interest, *when* there is little danger of drift due to uncontrolled causes, and *when* his equipment requires him to take one observation at a time, *then* he may wish to do his work in such an order as to obtain an early look at simple nonadditivities if they exist. Such "one-at-a-time" plans have little interest for two-factor problems, but begin to produce interesting and economical results when three factors are involved. See Daniel [1973].

When random error is small and one run at a time must be made, *but*, contrary to the case discussed above, the system may show drift over sets of runs, some orders of trials are much better than others. Such plans are discussed in Chapter 15.

3.6.3. When Is Randomization Useful?

Randomization becomes increasingly useful as we move away from the various situations described above. Thus (1) when the experimenters can produce only rough qualitative judgments of the magnitudes of effects (as multiples of σ, the standard deviation of single observations), (2) when effects of importance are of the order of σ, (3) when experiments are near terminal, (4) when lower-ranking or younger or even mediocre research workers must provide evidence that will convince their skeptical superiors, and (5) when serious, perhaps life-threatening, alternatives are under study in human experimentation, *then* randomization may be an essential part of the experimental design. In cases 2 and 5 especially, it is decisively important to secure statistically independent observations, since the experimenters are relying on the law of large numbers to reach an acceptably small standard error of each effect. Randomization brings us closer to statistical independence.

I have given more space to the discussion of valid nonrandomization than to the customary insistence on careful randomization. This has been necessary to counter the doctrinaire claims of some statisticians which have, I think, been responsible for repelling many research workers who might well gain by an occasional randomization.

I admire and recommend highly the pages on randomization in the classic book of Cochran and Cox [1957], especially Sections 1.13 and 1.14. There is nothing doctrinaire in these pages. It is true that there are no examples in this work of experimental data from any of the physical sciences. My own experiences, having been mainly in the latter areas, have surely been responsible for my differing emphasis.

The inventor of randomization, or at least its prime developer in experimental agriculture, was R. A. Fisher. Some of the claims made in his *Design of Experiments* [1953] are not accepted here. We quote (out of context, of course, as one aways quotes) from his page 9 of the sixth edition: "The chapters which follow are designed to illustrate the principles which are common to all experimentation, by means of examples chosen for the simplicity with which these principles are brought out." I do not find stated or illustrated the principles that are claimed to be common to all experimentation. Section 9 (bearing the title "Randomization: The Physical Basis of the Test") curiously does not mention randomization. Section 20 ("Validity of Randomization") does state that the simple precautions outlined supply absolute guarantee of the validity of experimental conclusions. It would have been sufficiently impressive if only an occasional improvement in validity had been claimed.

3.6.4. An Example

Federer [1955, pages 176 ff.] describes a 2^2 done in four randomized blocks of four. The responses are given as "per cents of bud-take success," and we are asked to assume that all per cents are based on the same number of grafts. We look at all the residuals by removing row and column effects from two-way Table 3.7.

Seeing no signs of serious heterogeneity, we call the residuals reflections of random error, but only with $(4 - 1)(4 - 1) - 1$ or eight d.f. We estimate the error variance by the mean square residual $= 823/8 = 102.9$, and the error standard deviation by $s = 10.14$. The standard error of the means of four observations will be $s/\sqrt{4} = 5.07$. The means for the four factorial experimental conditions are as follow:

$$
\begin{array}{ll}
(1) & 72 \\
a & 21 \\
b & 36 \\
ab & 19 \\
\end{array}
$$

Inspection tells me that calculation of effects is not likely to clarify anything; but if inspection does not tell the reader this, he should put the four

TABLE 3.7.

NAIK'S DATA, VIA FEDERER [1955, PAGE 177], FOR A 2^2 IN FOUR BLOCKS

Block	(1)	a	b	ab	$\sum_R$	Row Averages
I	64	23	30	15*	132	33
II	75	14	50	33	172	43
III	76	12	41	17	146	36
IV	73	33	25	10	141	35
$\sum_C$: 288		82	146	75	591	148
$\sum_C - 148$: 140		-66	-2	-73	-1	
Column Deviations: 35		-16	-1	-18		
Column Averages: 72		21	36	19		37

Residuals

$-4^\dagger$	6	-2	0*
-3	-13	8	8
5	-8	6	-1
3	14	-9	-7

* Observation missing; value inserted to give zero residual.

$\dagger$ Residual $= d_{ij} =$ observed value $-$ fitted value

$$= y_{ij} - Y_{ij} = y_{ij} - [y_{..} + (y_{i.} - y_{..}) + (y_{.j} - y_{..})]$$

$$= y_{ij} - y_{i.} - (y_{.j} - y_{..})$$

for example, $d_{11} = 64 - 33 - (72 - 37) = 64 - 33 - 35$

$$= -4.$$

means through Yates's algorithm and get, for M, $\hat{A}$, $\hat{B}$, and $\hat{AB}$, respectively, 37, -17.3, -9.3, and 8.5, all with standard error $s/\sqrt{16}$ or 2.54. The only simple finding seems to me to be that the response to (1) far exceeds the responses to the three other conditions. Nature has this time declined to respond in terms of main effects.

The main (but after all minor) point of this example is to provide a means of looking at the residuals which are reflections of the random fluctuations in the response. Later we will take a more severe position in judging the homogeneity of such residuals. Here it suffices to note that the residuals in the first column are not at all larger than the remainder, so that there is little point in a transformation of the data to gain some theoretical homogeneity.

The minor point (for us, major for the experimenters) of the example is that the partition into main effects and interaction has failed, and that this is not at all a rare outcome. The higher levels of the two factors that were varied have damaged "bud take" and have done so nearly uniformly.

The results of this small analysis can be put into "analysis of variance" format, as in Table 3.8.

TABLE 3.8.
ANALYSIS OF VARIANCE OF FEDERER'S REPLICATED 2^2

Source of Variation	Degrees of Freedom	Sums of Squares	Mean Squares	
Blocks	3	221	73.7	Nonsignificant
(1) vs. a, b, ab	1	6557	6557	Significant
Among a, b, ab	2	766	383	Nonsignificant
Residual	8*	823	103	
	14*	8367	by addition	
		8363	from data	

* One degree of freedom has been lost since only 15 (not 16) pieces of data are given in Table 3.7.

3.7. SUMMARY

The two-factor, two-level plan, the 2^2, is discussed in elementary detail, introducing the factorial representation with its various symbols, the definition of two-factor interactions, and Yates's "addition and subtraction algorithm" for computing all effects and interactions compactly. There is much homily about the interpretation of large 2fi.

The varying needs for randomization in engineering and other scientific experimentation are discussed at length. In many situations randomization is not required or is undesirable; in many others it is a desideratum; and in some it is a nearly absolute necessity. The need depends on the ratio of the effects to be detected to the error standard deviation *and* on the nearness of the program to final, costly decision.

APPENDIX 3.A

THE ANALYSIS OF VARIANCE IDENTITIES

A.1. One Set of Repeated Observations

Designating the observations as y_i ($i = 1, 2, \ldots, I$), and their mean as $y_.$, we write

(3.A.1) $$y_i \equiv y_. + (y_i - y_.).$$

This equation as it stands assumes nothing about the error distribution. But if the observations are a fair sample, their average estimates the population mean μ, and the $(y_i - y_.)$ are simply the residuals (not quite the random errors unless I is large, since they have variance $[(I - 1)/I]\sigma^2$).

An analogous relation holds for sums of squares, for, squaring both sides of (3.A.1),

$$y_i^2 = y_.^2 + (y_i - y_.)^2 + 2y_.(y_i - y_.),$$

and summing over i, we have

(3.A.2) $$\sum y_i^2 = \sum y_.^2 + \sum (y_i - y_.)^2 + 2y_. \sum (y_i - y_.)$$
$$= I y_.^2 + \sum (y_i - y_.)^2,$$

since

$$\sum (y_i - y_.) = 0.$$

This may be expressed in words as follows: The sum of the squares of I numbers is I times the square of their mean *plus* the sum of squares of the deviations of the numbers from their mean.

A.2.   Several Sets of Repeated Observations

Let i now designate the sets, taken presumably under different conditions, and j the replicates: $i = 1, 2, \ldots, I$, and $j = 1, 2, \ldots, J$, assumed to be the same number for all i. Calling the grand average $y_{..}$, we write:

$$y_{ij} \equiv y_{..} + (y_{i.} - y_{..}) + (y_{ij} - y_{i.})$$

(3.A.3)   or $$(y_{ij} - y_{..}) \equiv (y_{i.} - y_{..}) + (y_{ij} - y_{i.}),$$

which is obviously an identity for any set of identifiable $I \times J$ numbers. In words, we can say: The deviation of each of $IJ = N$ numbers from their grand average is equal to the deviation of its group average from the grand average plus its deviation from its group average.

Squaring, summing, and recognizing that sums of deviations from means are always 0, we have

(3.A.4) $$\sum_i \sum_j (y_{ij} - y_{..})^2 = J \sum_i (y_{i.} - y_{..})^2 + \sum_i \sum_j (y_{ij} - y_{i.})^2.$$

This is the "one-way analysis of variance identity" since the data are grouped only by i. It is conceivably useful for judging whether the groups really differ in their means more than would be expected on the evidence of the scatter of individual observations around their group averages.

The reader should take it as an exercise to derive the sum of squares identity for the one-way case when there are different numbers of observations, n_i, in each group.

A.3. Two-Way Layout, Unreplicated

The numbers are now cross-classified by $i = 1, 2, \ldots, I$ for rows, and $j = 1, 2, \ldots, J$ for columns. The identity becomes

$$(3.A.5) \quad y_{ij} - y_{..} \equiv (y_{i.} - y_{..}) + (y_{.j} - y_{..}) + (y_{ij} - y_{i.} - y_{.j} + y_{..}),$$

where the last term in parentheses is simply written in to force the identity. This partition can always be made. It will be most useful if either or both of the first two terms is large compared to the average of the third term, because then we are finding that the system that produced the numbers is responding largely additively to the row and column partitions.

Squaring, summing, and simplifying as before, we have

$$(3.A.6) \quad \sum_i \sum_j (y_{ij} - y_{..})^2 = J \sum_i (y_{i.} - y_{..})^2 + I \sum_j (y_{.j} - y_{..})^2$$
$$+ \sum_i \sum_j (y_{ij} - y_{i.} - y_{.j} + y_{..})^2,$$

which will be useful for producing comparable averages of the three terms.

A.4. Two-Way Layout With Replication

We now write

$$(3.A.7) \quad (y_{ijk} - y_{...}) = (y_{i..} - y_{...}) + (y_{.j.} - y_{...})$$
$$+ (y_{ij.} - y_{i..} - y_{.j.} + y_{...}) + (y_{ijk} - y_{ij.}),$$

where i and j have the same meanings as in Section A.3, and $k \, (k = 1, 2, \ldots, K)$ designates the repeated observations in each ij combination. The corresponding sum of squares identity is perhaps obvious:

$$(3.A.8) \quad \sum\sum\sum(y_{ijk} - y_{...})^2 = JK\sum(y_{i..} - y_{...})^2 + IK\sum(y_{.j.} - y_{...})^2$$
$$+ K\sum\sum(y_{ij} - y_{i..} - y_{.j.} + y_{...})^2$$
$$+ \sum\sum\sum(y_{ijk} - y_{ij.})^2.$$

Here, as in the cases given in Sections A.1 and A.2, we have a real measure of random error and so can hope to judge objectively the reality and relative importance of the first three terms on the right-hand side.

All of these cases, as well as many others to be given later, demonstrate decompositions of data into parts that may well be scientifically interesting, especially when I and J are fairly large. But when $I = J = 2$ there is little point in viewing the displacements of the two levels as two deviations of equal magnitudes and opposite signs from their mean. The reader probably knows, too, that for two observations, or means, a and b,

$$\sum_{i=1}^{2} \left(a - \frac{a + b}{2} \right)^2 = \frac{(a - b)^2}{2}.$$

The simple difference between the two means is more informative than its square since the former has a sign; the latter is always positive, and nothing is gained by squaring. We can express this in an equivalent way as follows: A difference can be compared with its standard error more intelligibly than a squared (and halved) difference with its variance.

We will usually find that classifications with three or more levels can also be broken down into simple comparisons and often that these comparisons are more informative than mean squares giving equal weight to all levels. Scheffé [1959] has shown how to judge objectively all comparisons (contrasts) in any balanced set of data when the standard assumptions are satisfied. See Brownlee [1965, Section 10.3, page 316].

CHAPTER 4

Two Factors, Each at Three Levels

4.1. INTRODUCTION

As we take one more plodding step toward enlarging our view of multi-factor plans, we come upon the 3×3, that is, the 3^2. This gives us the opportunity to discuss a few more aspects of experimentation and of the interpretation of experimental results. (The 3×2 is too small a step to take, so we leave it for incidental treatment later.)

Just as the 2^2 corresponds to the type of linear approximation familiar to applied mathematicians, physicists, and engineers who habitually substitute straight lines for other functions, so the 3^2 parallels second-order approxi-mation. But even when the three levels, or versions, of a factor are not points on an ordered continuum, there will be many situations in which industrial research workers will want to study all three at once. For example, there are three major sources of coffee beans and (at least) three degrees of roasting the beans, and any restriction on either condition forces a postponement of the time when usable results can be reported.

A properly conservative statistician (and who would want to retain a statistician who was not properly conservative?) may recommend that, if a 3^2 corresponds to the experimenter's needs, it be replicated, preferably more than twice, so as to get an unbiased and reasonably powerful test of significance. If data are inexpensive or can be acquired quickly *or* are to be used for a major decision, this requirement may be gladly met by the experimenter. There are many cases, however, where such a recommendation will result only in a scientist's going his own way, alone and unguided. Thus the national air-pollution data have been collected for only 3 years; they are reported grouped for small, medium, and large communities. They produce, then, a necessarily unreplicated 3^2. Examples are given later of carefully randomized, highly replicated sets of nine which proved to be practical, and indeed in each case constituted the only means to secure safe inferences of sufficient precision.

4.2. BOTH FACTORS HAVE NUMERICALLY SCALED LEVELS

Such factors are called continuous (meaning that their levels are potentially continuous) or quantitative, by most writers. The standard examples in engineering are temperatures, pressures, or concentrations of ingredients, when these are independent variables. Such factors are usually set at equally spaced levels. Following G. E. P. Box and his associates, we imagine the response surface above the x_1-x_2 plane to be a quadratic surface, represented analytically as:

$$(4.1) \qquad \eta = \mu + \beta_1 x_1 + \beta_2 x_2 + \beta_{11} x_1^2 + \beta_{22} x_2^2 + \beta_{12} x_1 x_2,$$

Adding a random error term, assumed here to be normal with mean 0 and variance σ^2, $N(0, \sigma^2)$, uncorrelated, so that the observed values, y_{ijk}, are

$$(4.2) \qquad\qquad y_{ijk} = \eta_{ij} + e_{ijk},$$

where $i = 1, 2, 3$ indexes the level of x_1, j indexes the level of x_2, and k indexes any replicates taken at x_{1i}, x_{2j}, we see that six parameters are required. There are then only three d.f. from the 3^2 for lack of fit.

The general second-order equation is hardly ever a model derived from subject-matter knowledge. Nearly always, as Box and Wilson [1951] suggested, it is equivalent to approximating theory by using the second-order terms in a Taylor series expansion of the true function about some point in a region judged to be approximable by these terms. It may happen that a real physical model is available, nonlinear in its parameters, and that this is approximable by a series whose coefficients are definite functions of the "real" parameters, but this case is not discussed here.

There is at least one serious defect in using a 3^2 to estimate a full quadratic equation; perhaps there are two. Both were pointed out and indeed rectified by Box, Youle, and J. S. Hunter [1954, 1955, 1957] long ago. The more serious one, to my mind, can best be indicated by a glance at the three fictitious 3^2's of Figure 4.1. The values in the cells are "responses." Each square has the same $\beta_{12}x_1x_2$, but the response at the center is different in each case. It is evident that the general shape and orientation of the response surfaces are heavily influenced by the center point. This suggests strongly that if any point can be replicated this one should be. Even triplication or quadruplication is desirable. It will be obvious to the moderately competent algebraist that Figure 4.1a has large negative β_{11} and β_{22}, whereas Figure 4.1c has large positive values for these terms. In Figure 4.1b both these terms are zero.

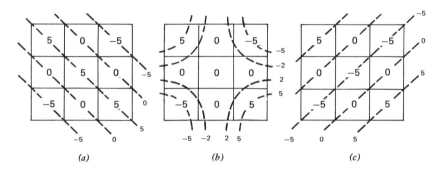

Figure 4.1   Three fictitious 3^2's, both independent variables continuous.

The second defect of the 3^2 as a second-order response surface design is the lack of radial symmetry about its center. It reaches further out from the center at its four vertex points than at its four midedge points. This produces contours of information about the center that are not circular. This handicap, although admittedly severe in plans in three or more factors, is at least minimal in the 3^2.

The important idea of second-order rotatability has been well explained many times, especially by Box [1954], Box and Youle [1955], and Box and Hunter [1957], as well as by Chew [1958], Cochran and Cox [1957], and Guttman, Wilks, and Hunter [1971]. It will not be discussed here. We regress to the old-fashioned 3^2.

Elementary analytical geometry suffices to derive the estimates of the co-efficients in (4.1). First, for a single, equally spaced x-variable, we look at Figure 4.2 and at the one-variable equation

(4.3) $$Y = a + bx + cx^2.$$

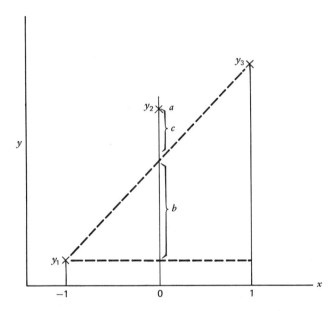

Figure 4.2 Geometric interpretation of coefficients in $Y = a + bx + cx^2$ when $x = -1, 0, +1$.

No least-squares fitting is required; only direct substitution of the three observed responses, y_1, y_2, and y_3, is needed in this equation to produce the three estimates:

(4.4) $$a = y_2, \qquad b = \tfrac{1}{2}(y_3 - y_1),$$
$$c = \tfrac{1}{2}(y_1 - 2y_2 + y_3) = \tfrac{1}{2}(y_1 + y_3) - y_2$$
$$= \tfrac{1}{2}[(y_1 - y_2) - (y_2 - y_3)].$$

The "linear" coefficient b is the slope of the line through the two extreme points. The quadratic coefficient c is the negative of the amount by which y_2 differs from the average of y_1 and y_3. It may be viewed alternatively as the difference in slope between the two line segments $y_2 - y_3$ and $y_1 - y_2$. The third way of writing this estimate, shown first in (4.4), gives the multipliers 1, -2, and 1 for y_1, y_2, and y_3 in simplest integer form.

The same forms apply to row averages (averages over x_2) and to column averages (over x_1), and all four estimates (b_1, b_2, b_{11}, b_{22} corresponding to the first four betas in (4.1) are orthogonal. Following Yates we will use the symbols A_L, B_L, A_Q, B_Q (the subscripts L and Q stand for *linear* and *quadratic*) instead of b_1, b_2, ...

Still intuitively, we note that the term $b_{12}x_1x_2$ can have effect only when both x_1 and x_2 are nonzero and so should take the same form here that it did for the 2^2, using only the four "corner" values of the 3^2. Direct check shows the estimate

$$b_{12} = \tfrac{1}{4}(y_9 - y_7 - y_3 + y_1)$$

to be orthogonal to the others.

We tabulate the estimates of the six effects [the β's in (4.1)]—five of them contrasts—in Table 4.1. The captions $\hat{A}_L \cdots \widehat{A_L B_L}$ are identical in meaning with their corresponding b's. The divisors are the sums of squares of the multipliers shown in each column. The whole table is, then, a coded "transformation matrix" for the 3^2.

TABLE 4.1.
ESTIMATES OF SIX EFFECT PARAMETERS FROM A 3^2

$A \equiv x_1$	$B \equiv x_2$	Obs.	b_0	$\hat{A}_L \equiv b_1$	$\hat{B}_L \equiv b_2$	$\hat{A}_Q \equiv b_{11}$	$\hat{B}_Q \equiv b_{22}$	$\widehat{A_L B_L} \equiv b_{12}$	Cell $(3,3)$
-1	-1	y_1	1	-1	-1	1	1	1	1
0	-1	y_2	1	0	-1	-2	1	0	1
1	-1	y_3	1	1	-1	1	1	-1	-2
-1	0	y_4	1	-1	0	1	-2	0	1
0	0	y_5	1	0	0	-2	-2	0	1
1	0	y_6	1	1	0	1	-2	0	-2
-1	1	y_7	1	-1	1	1	1	-1	-2
0	1	y_8	1	0	1	-2	1	0	-2
1	1	y_9	1	1	1	1	1	1	4
Divisor:			9	6	6	18	18	4	36

It is easy to verify that all six vectors are orthogonal. [The column headed "Cell $(3, 3)$" will be explained in Section 4.4.] We can compute a sum of squares for each column. Each will be the square of a single number (the inner product of the coefficient vector times the y-vector) *divided* by the given divisor. These values will be put into analysis of variance form in Table 4.3 in the next section. In my opinion it is more intelligible to show this partition arithmetically than algebraically.

The three d.f. for lack of fit to a full quadratic can be used to provide a "sum of squares for lack of fit." It is more in accord with my point of view to use them to estimate three more parameters. The classical parameters first given by Yates correspond to the three additional terms in the fitting equation:

$$b_{112}x_1^2x_2 + b_{122}x_1x_2^2 + b_{1122}x_1^2x_2^2,$$

chosen no doubt because they have easily derivable estimates orthogonal to all others and to each other. They do not appear to me to be plausible geometrically or analytically, but an effort will be made in Section 10.6 to give them two more intuitive interpretations.

When the three levels of each factor are evenly spaced, an extension of Yates's 2^n algorithm is available [Davies *et al.* 1959, pages 363–366] for the 3^2 and for the 3^m, $m > 2$. A further extension to factorial plans of the form 2^n3^m is given by Margolin [1967].

4.3. STANDARD COMPUTATIONS IN A 3^2

First we set down a standard computation that should precede any computation of a "sum of squares for interactions." Indeed we propose an inflexible rule: Do not try to interpret sums of squares of quantities whose summands you do not know. Refusal to inspect the items summed may lead (and, as we will abundantly document later, often has led) to the overlooking of a sensible subdivision that is informative. As a prime example, nearly all of the sum of squares may be in a simple single contrast or summand.

Since the arithmetic operations are simple, and their algebraic representations tedious, we start with a classical example from Cochran and Cox [1957, Ed. 2, page 164]. We reduce the arithmetic by subtracting the mean, 353, from each value in their Table 5.6. We then compute the row averages and column deviations, fitted values to the additive model, and residuals from the observed values, that is, the estimated interactions, as shown in Table 4.2. The rows correspond to three equally spaced levels of a phosphate fertilizer; the columns, to three equally spaced nitrogen levels. The responses are totals over 12 plots of the numbers of lettuce plants emerging.

TABLE 4.2.
STANDARD COMPUTATIONS ON COHRAN AND COX'S 3^2 [1957, PAGE 164] ON LETTUCE

Data			Data—353			$\sum_R$	Row Averages
449	413	326	96	60	−27	129	43
409	358	291	56	5	−62	−1	0
341	278	312	−12	−75	−41	−128	−43
		$\sum_C$:	140	−10	−130	0	
Column Deviations:			47	−3	−43		

Fitted Values				First Residuals		
90	40	0		6	20	−27
47	−3	−43	⇒	9	8	−19
4	−46	−86		−16	−29	45

SS residuals $= 4793$; SS $\left[\text{cell } (3, 3)\right] = \frac{9}{4}(45)^2 = 4556$; MS residual $= 1198$.
Maximum normed residual (MNR)* $= 45/4793^{1/2} = 0.650$; MNR $(0.05, 3 \times 3) = 0.648$.
$y^0(3, 3) = (−27 − 62)/2 + (−12 − 75)/2 − (96 + 60 + 56 + 5)/4 = −142.2$.

Revised Data			$\sum_R$	Row Averages	Second Residuals		
96	60	−27	129	43	−5	9	−4
56	5	−62	−1	0	−2	−3	4
−12	−75	−142	−229	−76	6	−7	0
$\sum_C$: 140	−10	−231	−101				
$\sum_C + 34$: 174	24	−197	1		New SS residual $= 235$		
Column					New MS residual $= 235/3$		
Deviations: 58	8	−66			$= 78.7$		

* Stefansky [1972] and Appendix 4.A.

39

TABLE 4.3.

TWO ANALYSES OF VARIANCE FOR THE 3 × 3 OF TABLE 4.2.

Source of Variation	Unrevised Data			One Cell [Cell (3, 3)] Revised		
	Degrees of Freedom	Sum of Squares	Mean Square	d.f.	SS	MS
Phosphate, P	2	11,008.67				
$P_{\text{Lin.}}$	1	11,008.17		1		21,360.67
$P_{\text{Quad.}}$	1	0.50		1		533.56
Nitrogen, N	2	12,200.00				
$N_{\text{Lin.}}$	1	12,150.00		1		22,940.17
$N_{\text{Quad.}}$	1	50.00		1		280.06
Interaction $P_L N_L$	1	2,209.00				
Remainder	3	4,791.33	1,198.00	3	235.10	78.4
Total	8	28,000.00		7	45,349.56	

4.4. ONE-CELL INTERACTION

On inspection of the nine residuals (first residuals) we spot the largest in cell (3, 3), but we need some objective way of judging this largeness. Ignoring the deplorable gaffe in the paper "Residuals in Factorials" [Daniel, 1961], we find an excellent answer in Stefansky [1972]. Her "maximum normed residual," $z^{(0)}$, is easy to compute; it is the ratio of the maximum residual to the square root of the residual sum of squares. For the present case we have a significance probability of roughly 0.05. See Appendix 4.A.

If these were engineering data, and if it could be done tactfully, we would ask the experimenters whether by any chance the reported 312 (coded to -41 in Table 4.2) could have been misrecorded by 100. But this can hardly be done, even tactfully, for ancient data in another field. We try instead to remove the disturbing effects of this value, to see whether anything else is afoot. We replace the offending value by another, y^0, which is computed to give a zero residual, and then carry through the whole computation again. This is done at the bottom of Table 4.2. The drastic effect on all residuals is apparent.

The replacement value y^0 is given by the formula:

$$(4.5) \qquad y_{ij}^0 = y_{i.}' + y_{.j}' - y_{..}',$$

where $y_{i.}'$ means the average of all *other* values in row i,

$y_{.j}'$ means the average of all *other* values in column j,

$y_{..}'$ means the average of all values in neither.

This formula is exactly equivalent to the usual "missing value formula" for randomized blocks, but seems to me more intelligible, easier to remember, and simpler to compute. It holds for any size of two-way layout.

A simple check on any residual in a 3^2 is provided by the identity

(4.6) $d_{ij} = \frac{1}{3}(y_{ij} - \sum y_{i'j} - \sum y_{ij'})$,

provided that

$$\sum\sum y_{ij} = 0; \qquad \text{Var}(d_{ij}) = \frac{4}{9}\sigma^2.$$

For the present case $d_{33} = \frac{1}{3}[-41 - (-27 - 62 - 12 - 75)] = 45.0$.

Our sums of squares (SS's) are just 12 times those of Cochran and Cox since we have so far ignored the fact that each response is the sum of 12 observations. On a plot basis our revised residual mean square (RMS) is $78.7/12 = 6.56$. This is significantly *smaller* ($P \doteq 0.05$) than the error mean square (MS) of 59.0 given on page 166 of the reference.

We have an example, then, of a *one-cell interaction*. These interactions are, in my experience, the commonest of all forms of nonadditivity, and for "quantitative-level" factors they occur most frequently in a corner cell. We can derive the coefficients for the "one-cell contrast" most directly by simply placing a single disturbance in one cell, and then following it through the computation corresponding to Table 4.2. To get minimal integers we put a 9 in one cell. Thus we have

				$\sum_R$ Averages		Residuals		
	9	0	0	9	3	4	−2	−2
	0	0	0	0	0	−2	1	1
	0	0	0	0	0	−2	1	1
$\sum_C$:	9	0	0	9				
$\sum_C - 3$:	6	−3	−3					
Column Deviations:	2	−1	−1					

We have used cell (1, 1) above, but of course the same pattern can emerge from any cell. The simplest way to remember the pattern is to place a 4 in the offending cell and then fill out the rest, forcing rows and columns to sum to zero.

The reader may wish to verify that this pattern, or some constant multiple of it, will emerge from any 3×3 table of additive data when *one* value is perturbed by any amount. A constructed example follows. If the data also contain random error, the pattern will of course be more or less obscured.

$$\begin{array}{ccc} 28 & 36 & 47 \\ 58 & 48 & 59 \\ 85 & 75 & 86 \end{array}$$

Although the contrast just derived for a one-cell interaction is correct and can be used with the original data, there is little point in its use with the residuals. Naming the residuals,

$$\begin{array}{ccc} p & q & r \\ s & t & u \\ v & w & x, \end{array}$$

we see that $-2q - 2r = 2p$, $-2s - 2v = 2p$, and $(t + u) + (w + x) = -s - v = p$. So our integer contrast is identically $9p$! The corresponding SS is $(9p)^2/36 = \frac{9}{4}p^2$. For the present case we have that the SS [interaction in cell $(3, 3)$] $= \frac{9}{4}(45^2) = 4556$, exactly as would be found by subtracting the SS new residuals from the original SS residuals, $4793 - 237 = 4556$. The computation of "second residuals" gives more detail on what is left, even though containing only three degrees of freedom.

We now see the most direct way to calculate the replacement value y_{ij}^0, and, once the first residuals have been computed, all the second residuals. For the 3^2: $y_{ij}^0 = y_{ij} - \frac{9}{4}d_{ij}$. For the present case $y_{3,3}^0 = -41 - \frac{9}{4}45 = -142$.

The pattern of *differences* between second and first residuals is just the negative of the familiar pattern, hence here,

$$\begin{array}{ccc} -1 & -1 & 2 \\ -1 & -1 & 2 \\ 2 & 2 & -4 \end{array}$$

each coefficient scaled by $45/4$, or 11.25. Thus the new $d_{11} = 6 - 11.25 = -5.25$.

The weakness of the standard representation of interactions in a factorial plan

(4.7) $$\eta_{ij} = \mu + \alpha_i + \beta_j + \gamma_{ij}$$

is nearly at its maximum in the 3^2. The γ_{ij} are defined by subtraction of "main effects" from η_{ij}, but as the little table above with one 9 in it makes clear, if one value is off from a clear pattern set by the rest of the table, then all consistent main effects are biased, as is the mean, and the rest of the disturbance gets spread throughout the table as shown. These $\hat{\gamma}_{ij}$ do not estimate anything real, but by good luck, the largest of them can be used to recover the real discrepancy which is $\frac{9}{4} \times 45 = 101.25$. In general, in an

$R \times C$ table, this correction factor will be $RC/(R - 1)(C - 1)$. See Table 8.1b. If this value is actually an error, and not a permanent interaction, then we do the experimenter no favor by reporting biased main effects and a spurious linear-by-linear interaction.

4.5. SIMPLER COMPUTATION AND INTERPRETATION OF $A_L B_Q$, $A_Q B_L$, AND $A_Q B_Q$

If the data analyst has a completed 3^2 before him (A and B continuous) and a good estimate of σ, he may want to see explicitly the three higher-order coefficients, b_{122}, b_{112} and b_{1122} (alternatively, $A_L B_Q$, etc.) There are two ways of computing these that are simpler than the usual ones, originally given by Yates [1937], shown at the top of Table 4.4.

TABLE 4.4.

COMPUTATION OF $A_L B_L$, $A_L B_Q$, ETC., FROM RESIDUALS IN A 3^2

	p	q	r	s	t	u	v	w	x
$12A_Q B_L$	-1		1	2		-2	-1		1
$12A_L B_Q$	-1	2	-1				1	-2	1
$36A_Q B_Q$	1	-2	1	-2	4	-2	1	-2	1
$4A_L B_L$	1		-1				-1		1
$4A_Q B_L$	-1		-1				1		1
$4A_L B_Q$	-1		1				-1		1
$4A_Q B_Q$	1		1				1		1
$4A_L B_L$	1		-1				-1		1
$4A_Q B_L$		1						-1	
$4A_L B_Q$				1		-1			
$4A_Q B_Q$					1				

Since there are but four d.f. in the nine residuals, we choose four of the latter, p, r, v, x, and represent the four parameter estimates by the familiar contrasts for the 2^2 among these four and by their sum. Because of the restraints on the nine residuals, $q = -p - r$, etc., it is not necessary or even sensible to use all nine in computation. Going further, we see in the lowest panel of the table that we can replace $-p - r$ by q, etc., and so represent $4A_L B_Q$, for example, by $(s - u)$, and $4A_Q B_Q$ by t!

On soberer thought these quantities make good sense as aspects of lack of fit. Take $4A_L B_Q$, for example. This will be large when the four corner residuals have the following signs:

$$-p \qquad +r$$
$$(+s) \qquad (-u)$$
$$-v \qquad +x.$$

This can happen only when there is a sort of reversal from left to right in the response surface, since if p and v are negative, then s must be large and positive, while u must be large and negative. Thus the apparent curvature of the response away from the fitted quadratic surface at the lowest level of B is the reverse of that at its highest level. This can be visually represented as a bit of

overlaid on an otherwise quadratic and additive surface.

The usual (Yates) contrast for $A_Q B_Q$ is given by the coefficients

$$\begin{array}{rrr} 1 & -2 & 1 \\ -2 & 4 & -2 \\ 1 & -2 & 1, \end{array}$$

and this is, by analogy with our earlier one-called interaction, identical with $9t$ (t is the residual in the center cell). Some sombrero-type deviation from additivity is indicated:

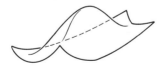

When a three-level factor is continuous but not equally spaced, there is also a simple way of writing orthogonal L and Q contrasts and SS's. We could call the three levels u, v, and w, (these have nothing to do with the residuals u, v, w) but things are easier to remember if we rescale the levels to 0, 1, and d, where $d > 1$ but is otherwise unrestricted. We give the formulas in both codings:

Levels$_1$	L_1	Q_1	Levels$_2$*	L_2	Q_2
u	$2u - v - w$	$v - w$	0	$-1 - d$	$1 - d$
v	$-u + 2v - w$	$w - u$	1	$2 - d$	d
w	$-u - v + 2w$	$u - v$	d	$-1 + 2d$	-1
Divisors for SS:				$6(1 - d + d^2)$	$2(1 - d + d^2)$

* Exactly equivalent to L_1, since subtracting u from each of Levels$_1$ gives 0, $v - u$, and $w - u$; then division by $v - w$ gives 0, 1, $(w - u)/(v - u)$. I have set the latter ratio equal to d.

The mnemonic is: Each term in Q_2 is the difference in cyclical order between the other two levels. I advise the reader not to try to get a comparably simple formula for four unevenly spaced levels.

4.6. TUKEY'S TEST FOR MULTIPLICATIVE NONADDITIVITY

As is explained most simply in Scheffé [1958, page 130], the interactions, with $(R - 1)(C - 1)$ degrees of freedom (for us here 4 d.f.), *may* be representable as

$$\gamma_{ij} = G\alpha_i\beta_j,$$

where G is a constant, and so a single degree of freedom for interaction. Scheffé shows that, if the γ_{ij} are representable by a second-degree polynomial in the row and column parameters, it must be of the form just given. Tukey showed [1949] that G can be estimated by

$$\tilde{G} = \frac{\sum\sum\hat{\alpha}_i\hat{\beta}_j y_{ij}}{\sum\hat{\alpha}_i^2 \cdot \sum\hat{\beta}_j^2} = \frac{P}{Q},$$

and that SS $(G) = P^2/Q$.

These are most simply computed as shown in Snedecor [1965, Ed. 5, seventh printing, pages 321–323]. We repeat the calculation in Table 4.5 for the 3^2 under discussion, and we regret that the test does not reach the 0.05 level. It should be noted that the $\hat{\gamma}_{ij}$ can be used in place of the y_{ij}. The estimated "error sum of squares" for testing nonaddivity is found to be 2817, which far exceeds the 236 found by our brutally ad hoc removal of $y_{3,3}$. The $\hat{\gamma}_{ij}$ estimated from $G\hat{\alpha}_i\hat{\beta}_j$ do not suffice to remove a large part of the residual SS, as they will when this relation holds strongly, and when, therefore, the $\hat{\alpha}_i$ and the $\hat{\beta}_j$ are sufficiently large to permit its accurate estimation.

TABLE 4.5.*

TUKEY'S TEST FOR REMOVABLE NONADDITIVITY. DATA FROM COCHRAN AND COX
[1957, PAGE 164]

Data			$\sum_R$	Row Averages	d_i Dev.	p_i	$d_i p_i$	$p_i - \bar{p}^*$
449	413	326	1188	396	43	5397	232,071	1377
409	358	291	1058	353	0	5227	0	1207
341	278	312	931	310	$-43^†$	1436	$-61,748$	-2584
$\sum_C$: 1199	1049	929	3177		0	12,060	170,323	
Column Averages: 400	350	310	353			4,020		
Column Deviations d_j: 46	-3	$-43^†$	0					

$p_i = \sum X_{ij}d_j$; $p_1 = (449 \times 46) - (413 \times 3) - (326 \times 43) = 5397$.
$\sum\sum\hat{\alpha}_i\hat{\beta}_j y_{ij} = P = \sum d_i p_i = 170,323$; $\sum d_i^2 = 3698$; $\sum d_j^2 = 3974$.
SS (nonadditivity) $= P^2/\sum d_i^2 \cdot \sum d_j^2 = 1974$; $G = P/Q = 0.01159$.
SS (testing) $=$ Residual SS $-$ SS (nonadditivity) $= 4791 - 1940 = 2817$.
MS (testing) $= 2817/3 = 939$; F-ratio (nonadditivity) $= 1940/939 = 2.07$, n.s.

* Var $(p_i - \bar{p}) = \frac{2}{3}\sum d_j^2$; MS (testing) $= \frac{2}{3} \times 3974 \times 939 = 2.488 \times 10^6$; $s(p_i - \bar{p}) = 1577$.
† Force $\sum d_j = \sum d_i = 0$.

4.7. AN EYEBALL TEST FOR INTERACTION

When additivity holds, differences between observations in adjacent rows should be tolerably constant. The writing down of these differences usually suffices to spot a single nonadditive cell, and does so in these data. Thus, even for the uncoded data,

			Row Differences		
449	413	326	40	55	35
409	358	291	68	80	-21
341	278	312			

It is clear with no knowledge of σ at all that the last difference, $y_{2,3} - y_{3,3}$, is far from its mates, 68 and 80. The "tolerable" agreement of the three differences from rows 1 and 2 tells us that $y_{3,3}$ is the culprit, not $y_{2,3}$. The objective statistician will wince at the word *tolerable*, but it is our experience that when this test fails, objective tests do too. On the contrary, we will see several cases in which we can spot the most aberrant cell or cells by the row difference pattern, but cannot prove a real discrepancy by the maximum normed residual (MNR) method.

4.8. WHAT IS THE ANSWER? (WHAT IS THE QUESTION?)

There is a single failure of additivity in the data of Table 4.2. If we bar the invidious suspicion that the value at (3, 3) is a misprint, in error by 100, the discrepancy is in such a direction that the adverse effects on lettuce plant emergence of increasing phosphate and nitrogen are not so great at the extreme condition (highest P and N) as the other data would lead us to expect. This can hardly be an agriculturally important finding, however, since the general conclusion already visible is that the least amounts of phosphate and of nitrogen are most favorable. Presumably, if fertilizer is needed on this variety of lettuce, it should be applied after emergence.

4.9. AN UNREPLICATED 3^2 ON AIR-POLLUTION DATA

The August 1972 (Third Annual) Report of the Council on Environmental Quality gives on page 9 a table of "extreme value indexes" or EVI's, which are measures of the worst air conditions observed for each of three years (columns), and for communities of three sizes, $< 10^5$, $< 4 \times 10^5$, and $> 4 \times 10^5$ (rows). Table 4.6 gives these values, multiplied by 100, followed by standard computations.

The very satisfactory partition given in the analysis of variance table—with both row and column effects highly significant against a 4 d.f. error estimate—tempts us to let well enough alone. But a rather peculiar improvement is still possible. We have carried through, but do not show here, the 1 d.f. test for removable nonadditivity. Its F-value is 6.94, while $F(.05)$ is 10.1, and $F(.1)$ is 5.5. However, since there appears to be a positive trend of p_i with $\hat{\alpha}_i$, we have taken the liberty of extracting the square root of each EVI and carrying through the standard computations once more. We now get F-values of 379. and 299. and significance probabilities below .0005. Since the coefficient of variation of the transformed variable is 1.46%, the imputed coefficient of variation of the back-transformed EVI would be about 3%.

The EVI's are themselves the square roots of sums of squares of concentrations of three pollutants. We cannot imagine why the square roots should more nearly fit an additive model, but we have made the facts known to the proper authorities.

TABLE 4.6.

$100 \times$ EVI (EXTREME VALUE INDEXES) OF AIR POLLUTION, 1968–1970, FOR
THREE SIZES OF COMMUNITIES

Cities	1968	1969	1970	Row Deviations	Residuals		
Small	1035	768	641	96	-12	13	-2
Medium	661	410	334	-251	-39	2	38
Large	1156	799	666	155	⑤⓪	-15	-36
Column Deviations:	232	-60	-172	$719 = y_{..}$			

MNR $= 50/7306^{1/2} = 0.585$; $z^{(0)}(0.2, 3 \times 3) = 0.620$, so $P > .2$.

Analysis of Variance

Source	Degrees of Freedom	Sums of Squares	Mean Squares	F	P
Sizes (R)	2	287,723	143,861	79	$<.001$
Years (C)	2	260,560	130,280	71	$<.001$
Residuals	4	7,306	1,827	$s = 42.8$; decoded, 0.43.	
				CV $= 43/719 = 6\%$.	

4.10.   THE 3^2 WITH BOTH FACTORS DISCONTINUOUS

We take Cochran and Cox's third 3^2, given in their Section 5.29, pages 170–175, which was actually replicated four times so that an error estimate with 24 d.f. is available. The nine treatment means are given in Table 4.7, followed by a table of residuals from the usual additive model.

We notice the largest residual at $(1, 3)$, but we do not judge its importance by the maximum normed residual because we have a better estimate of σ. We use, rather, the Studentized extreme deviate t' [Pearson and Hartley 1954, page 173], which allows for the error degrees of freedom *and* for the number of values of which $d_{1,3}$ is the extreme. Since the residuals in a 3^2 have variance $\frac{4}{9}\sigma^2$, we estimate $s(d_{1,3})$ as $\frac{2}{3} \times 2.58 = 1.72.$, and so our $t' = 5.9/1.72 = 3.43$. Since the .01 value from the table is 3.22, we appear to have an excessive d_{ij}.

As for the earlier example, it is easier to compute the SS due to the single residual by $(9/400)59^2 = 78.3$ rather than by the elaborate contrast with coefficient 4 in cell $(1, 3)$, balanced symmetrically in the other eight cells.

TABLE 4.7.

ABBREVIATED COMPUTATIONS FOR COCHRAN AND COX'S 3^2
ON COMPOST

Data			(Data $-$ 70.5) $\times$ 10		
53.6	56.8	67.0	-169	-137	-35
80.8	82.3	80.5	103	118	100
74.3	69.1	70.0	38	-14	-5

First Residuals			SS (residuals)/100 = 97.02.		
-46	-12	59	$s(d_{ij}) = \frac{2}{3}s(y) = \frac{2}{3} \times 2.58^* = 1.71.$		
5	22	-27	Studentized extreme deviate = 5.9/1.72		
41	-9	-31			= 3.43.
			$t'(0.01, 9, 24) = 3.22; P < .01.$		

Tukey's 1 d.f. for nonadditivity gives $F = 4.4$, nonsignificant.
SS $(d_{1,3}) = \frac{9}{4} \times 5.9^2 = 78.3.$
Remaining SS (residual) $= 97.0 - 78.3 = 18.7.$

* Residual MS $= 18.7/3 = 6.2$; Error MS from 24 d.f. $= 6.56$

4.11.   THE 3^2 WITH ONE FACTOR CONTINUOUS, ONE DISCRETE-LEVELED

Our data are again taken from Cochran and Cox [1957, pages 169–170]. The first paragraph of their general comment on interpretation of the analysis of variance is too valuable to paraphrase.

The separation of the treatment comparisons into main effects and interactions is a convenient and powerful method of analysis in cases where interactions are small relative to main effects. When interactions are large, this analysis must be supplemented by a detailed examination of the nature of the interactions. It may, in fact, be found that an analysis into main effects and interactions is not suited to the data at hand. There is sometimes a tendency to apply the factorial method of analysis mechanically without considering whether it is suitable or not, and also a tendency to rely too much on the initial analysis of variance alone when writing a summary of the results.

The experiment involved the response to three levels of nitrogen fertilizer (150, 210, 270 lb/acre) by three varieties of sugar cane. The 3^2 was done in four replications, and the error MS with 24 d.f. was 43.91. The corresponding coefficient of variation was about 5%.

TABLE 4.8.
DATA FROM COCHRAN AND COX'S 3^2 ON SUGAR CANE

	n_1	n_2	n_3	Row Differences		
v_1	266.1	275.9	303.8	20.3	25.7	22.1
v_2	245.8	250.2	281.7	-28.6	-7.9	50.1
v_3	274.4	258.1	231.6			

The row differences in Table 4.8 show instantly that v_1 and v_2 are nearly "parallel," and that v_3 is entirely different. A simple plot of the level of N versus $\bar{y}$ for each variety shows the same thing. We decline, therefore, to report on an analysis of variance of all three varieties since we know that it will show large $V \times N$ interaction, all due to v_3. The data can, as in many cases, be partly interpreted before a routine analysis of variance. We subdivide our analysis even in the first round, then, taking v_1 and v_2 without v_3. This is shown in Table 4.9. We partition the three v_3 means separately, but use the pooled error to test the partition.

It appears, then, that the following hold:

1. There was a consistent difference between v_1 and v_2 of $68.1/12 = 5.67$ tons of cane per plot.
2. There was an upward roughly linear trend of yield with increasing N for varieties 1 and 2.
3. There was an almost exactly linear downward trend of yield with N for v_3.

There are oddities in items 1 and 3. The parallelism of v_1 and v_2 is improbably close, *and* for v_3 the linearity of yield with N is too exactly linear! Both are significantly smaller than the error MS, given as 43.91. The mystery is only deepened by the authors' footnote on page 170, which points out that much of variety 3 ripened earlier than the other two but was left on the ground until harvest. Not having the actual data from which the error was computed, we terminate our analysis.

The attentive reader will notice that, if we had gone ahead with our standard computations, without taking account of the disparate variety, no great harm would have befallen us beyond wasted time. The rounded residuals are as follows:

	n_1	n_2	n_3
v_1	-13	-2	15
v_2	-10	-5	15
v_3	23	7	-30

These results would have told us that v_1 and v_2 are closely alike, whereas v_3 is quite different.

TABLE 4.9.
COMPUTATIONS AND ANALYSIS OF VARIANCE FOR 3^2 ON SUGAR CANE

	Data Coded by -270.6			$\sum_R$	Data for V_3	
v_1	-4.5	5.3	33.2	34.0	n_1	19.7
v_2	-24.8	-20.4	11.1	-34.1	n_2	3.4
					n_3	-23.1
$\sum_c$:	-29.3	-15.1	44.3	-0.1		
$v_1 - v_2$:	20.3	25.7	22.1	68.1		
Differences -22.7:	-2.4	3.0	-0.6			

SS $(v_1 - v_2) = (34.1 + 34.0)^2/6 \times 4 = 193.32.$ SS $(v_3) = 233.31.$
SS $[(N)v_1, v_2] = 3048.99/2 \times 4 = 381.12.$ SS (v_3, linear)
SS $(v_1, v_2 \times N) = (2.4^2 + 3.0^2 + 0.6^2)/8$ $\quad = (19.7 + 23.1)^2/2 \times 4$
$\quad = 15.12/8 = 1.89.$ $\quad = 228.98.$

ANALYSIS OF VARIANCE

Source	Degrees of Freedom	Sum of Squares	Mean Square	F	P
$v_1 - v_2$	1	193.32	193.3	4.4	.05
$N_L(v_1, v_2)$	1	338.56	338.6	7.7	.05
$N_Q(v_1, v_2)$	1	42.5	42.5	1.0	.5
$(v_1 - v_2) \times N$	2	1.89	0.945	0.0215	.9995
$v_3 \times N$, linear	1	228.98	229.0	5.22	.05
$v_3 \times N$, quad.	1	4.33	4.3	0.099	.9995
Error	24		43.91		

4.12. SUMMARY

Our analysis of 3×3 tables of data differs from the analyses of others mainly in the detailed treatment of interactions (nonadditivity of row and column effects, residuals, $\hat{y}_{ij}$).

Simplification of "linear by quadratic" and of "quadratic by quadratic" interaction terms has given them new meanings.

The commonest interactions appear to be either in one cell or in one row (or column). (This finding is extended in Chapter 8 for larger arrays.)

The numerical methods of analysis of variance remain valuable, but rather different partitions are used.

APPENDIX 4.A

CRITICAL VALUES OF THE
MAXIMUM NORMED RESIDUAL (MNR)*

TABLE 4.A.1.
STEFANSKY'S TABLE 6.1: CRITICAL VALUES OF THE MNR AT
LEVEL $\alpha = 0.01$

C \ R	3	4	5	6	7	8	9
3	.660	.675	.664	.646	.626	.606	.587
4		.665	.640	.613	.588	.565	.544
5			.608	.578	.551	.527	.506
6				.546	.519	.495	.475
7					.492	.469	.449
8						.446	.426
9							.407

TABLE 4.A.2.
STEFANSKY'S TABLE 6.2: CRITICAL VALUES OF THE MNR AT
LEVEL $\alpha = 0.05$

C \ R	3	4	5	6	7	8	9
3	.648	.645	.624	.600	.577	.555	.535
4		.621	.590	.561	.535	.513	.493
5			.555	.525	.499	.477	.457
6				.495	.469	.447	.428
7					.444	.423	.405
8						.402	.385
9							.368

* Reprinted by permission from W. Stefansky [1972a].

CHAPTER 5

Unreplicated Three-Factor, Two-Level Experiments

5.1. WHEN TO USE THE 2^3

The experimenter hearing of factorial experiments for the first time may feel that he should start with the simpler, more manageable plans. But if he does this, he will be postponing the time when he can easily see their advantages. Sixteen-run plans are usually more than twice as informative *per run* as eight-run plans. My advice to the experimenter considering a 2^3 is, then: If you really have three and only three factors that it makes sense to vary, if your time and money budgets are so restricted that eight runs will consume quite a large part of your effort, if you are quite sure that differences larger than two standard deviations are all that you are interested in, and if you are quite sure that you can choose levels for your factors so that all eight combinations will be operable, *then* the 2^3 is the one to use.

5.2.   A REAL 2³

The data given in Table 5.1 come from the early stages of a study of the effects of three well-known factors—time of stirring A, temperature B, and pressure C—on the thickening time of a certain type of cement. The response y was the time in minutes required to reach a certain degree of hardness. The exact specification of the two levels of each factor and the name of the particular cement type, although crucial for the experimenter, are not important here. The error standard deviation for single runs was known to be about 12 minutes. The runs were made in random order but are presented in the table in "standard" order.

TABLE 5.1.
DATA FOR AN UNREPLICATED 2³ IN
STANDARD ORDER: SD $\doteq$ 12

(1)	297
a	300
b	106
ab	131
c	177
ac	178
bc	76
abc	109

Most obviously the *simple A*-effects are small and rather uneven. They are as follows: $300 - 297 = 3, 131 - 106 = 25, 178 - 177 = 1$, and $109 - 76 = 33$. Taken individually, these results are hardly striking, since their standard error is about $12\sqrt{2}$ or 17.

Second, the results at high B are all much lower than their counterparts at low B. Thus with no computation at all we see that

$$b - (1) = -191,$$
$$ab - a = -169,$$
$$bc - c = -101,$$
$$abc - ac = -69.$$

It is clear that all these differences are real and that they do not agree very well with each other.

Finally, looking at the simple C-effects, we find $-120, -122, -30$, and -22. The C-differences are large at low B and small at high B. Some readers will see immediately that, when the B-effects are larger at low C than at high, the C-effects at low B must be larger than those at high B.

5.3. YATES'S TABLE OF SIGNS

Table 3.4 shows the four ways in which the four results of a 2^2 are combined to give the four (regression) coefficients of the factorial representation, and Table 3.5 shows the corresponding computation. Both the table of signs and the computational method are due to Yates. We now extend the table of signs and the corresponding computation to the 2^3.

The 2^2 in the upper left corner of Table 5.2 is seen to be made up of four squares, three of the form $^{+\,-}_{+\,+}$, and one in the upper right of the reverse form, $^{-\,+}_{-\,-}$. The 2^3 table of signs will require eight rows and columns. The eight parameters are symbolized in the column headings of Table 5.2. The signs in the table may be written down directly as three squares identical with the 2^2 table, and one in the upper right the same with all signs reversed. The dotted lines in Table 5.2 are meant to guide the eye to the component 2^1 and 2^2 tables. In this table we have dropped the 1's and retained only the signs. The letter T is used to indicate "Total"—formerly "Sum."

Table 5.2 is the "transformation matrix" that shows how the eight responses are to be handled to get their sum T and the seven contrast-sums $(A), \ldots, (ABC)$. If the latter are divided by 4, we have the "effects and interactions" as these are usually defined; dividing by 8, we get the "regression coefficients" of the factorial representation. The reader will remember that this expresses each result as the sum of a set of displacements due to different combinations of the factor levels, up or down from the grand average. Each column of Table 5.2 contains the ordered elements of an eight-dimensional vector. All eight are pairwise orthogonal (the Greek word for "perpendicular"). The definition of orthogonality of two column vectors is that the sum of the products of corresponding elements in the two columns is zero.

TABLE 5.2.
FACTORIAL TRANSFORMATION MATRIX FOR THE 2^3

Spec.	Obs.	T	A	B	AB	C	AC	BC	ABC
(1)	y_1	+	−	−	+	−	+	+	−
a	y_2	+	+	−	−	−	−	+	+
b	y_3	+	−	+	−	−	+	−	+
ab	y_4	+	+	+	+	−	−	−	−
c	y_5	+	−	−	+	+	−	−	+
ac	y_6	+	+	−	−	+	+	−	−
bc	y_7	+	−	+	−	+	−	+	−
abc	7_8	+	+	+	+	+	+	+	+

5.4. YATES'S ALGORITHM FOR THE 2^3

The reader will have guessed that the data from a 2^3 can be made to yield all eight "effects" by extending the algorithm of Yates, given in Section 3.3 for the 2^2, through *three* columns. This is done in Table 5.3. The algorithm works of course for any 2^p set of data. It has been programmed for all computers for $p = 4, 5, 6, 7$, but a single 2^4 computation by hand takes only from 10 minutes to $\frac{1}{2}$ hour, depending on the number of figures in each response. It is rare for three figures to be required, unheard of for four.

TABLE 5.3.
YATES'S ALGORITHM APPLIED TO A 2^3

Spec.	(0)	(1)	(2)	(3)	(3) ÷ 8	Name
(1)	297	597	834	1374	172	Average
a	300	237	540	62	8	$\hat{A}$ Time of stirring
b	106	355	28	-530	-66	$\hat{B}$ Temperature
ab	131	185	34	54	7	$\hat{A}\hat{B}$
c	177	3	-360	-294	-37	$\hat{C}$ Pressure
ac	178	25	-170	6	1	$\hat{A}\hat{C}$
bc	76	1	22	190	24	$\hat{B}\hat{C}$
abc	109	33	32	10	1	$A\hat{B}\hat{C}$

It is instructive to remove an eye average, say 170, from each of the eight observations, and then to repeat the calculation. The reader will find too that rounding the numbers to the nearest *10* before computation yields results that closely resemble those given. It is not always safe to round a number to a major fraction of its standard deviation. Cochran and Cox [1957] recommend rounding to not more than one quarter of a standard deviation. This rule is safe whatever the size of the collection of data under study, but it will be found to be increasingly conservative as $N = 2^p$ increases.

5.5. FIRST INTERPRETATION OF THE 2^3

Knowing that the error standard deviation of a single run is 12, we estimate the standard error (abbreviated as SE hereafter) of the contrast-sums in column (3) of Table 5.3 to be $12\sqrt{8}$ or 34. Alternatively, the estimated SE of the regression coefficients in the next column is $12/\sqrt{8}$ or 4.2. On direct inspection of the magnitudes of all seven effects, B and C and BC appear to be real.

TABLE 5.4.

$B \times C$ TABLE FOR THE DATA OF THE 2^3

		B				B	
		—	b			—	b
C	—	297 300	106 131		C	298	118
	c	177 178	76 109			178	92

These judgments confirm our first inspection of the data. Table 5.4 shows at the left the eight values of observations arranged in a "$B \times C$" two-way table. These pairs are averaged to give the form shown on the right.

It is clear that the effect of higher temperature B, in decreasing the setting time is much larger at low pressure (low C) than at higher pressure.

5.6.   REVERSE YATES'S ALGORITHM

The fitted values just found can be computed directly by the "reverse Yates" used in Section 3.3. Although there is no gain in economy in the present case, the work is carried through in Table 5.5 for practice.

TABLE 5.5.

REVERSE YATES'S ALGORITHM APPLIED TO THE REAL EFFECTS TO GIVE THE FITTED VALUES

	(0)	(1)	(2)	(2) ÷ 8 = Y	Name	y
(BC)	190	−104	740	92	$\hat{bc}$	76, 109
(C)	−294	844	1420	178	$\hat{c}$	177, 178
(B)	−530	−484	948	118	$\hat{b}$	131, 106
T	1374	1904	2388	298	$(\hat{1})$	297, 300

5.7.   INTERPRETATION WITH ONE FACTOR DISCONTINUOUS

If factor B were one with discontinuous versions or levels while C was a "continuous" variable, it would not make good sense to report the fitting equation as the last column of Table 5.3 implies, namely, as

$$(5.1) \qquad Y = 172 - 66x_2 - 37x_3 + 24x_2x_3,$$

where $x_2 = -1$ at low B,
     $= +1$ at high B;
   $x_3 = -1$ at low C,
     $= +1$ at high C.

Since the interaction term has no obvious interpretation under these conditions, it would be simpler to give *two* linear equations, one holding at the low level of B, the other at high B:

(5.2) $$Y_{B-} = 238 - 61x_3,$$

(5.3) $$Y_{B+} = 106 - 13x_3.$$

Each of these equations is derived (if that is not too pretentious a word) from the relevant half of the 2^3 by setting $x_2 = -1$ or $+1$ in (5.1). The reader should carry through these calculations, which require a small piece of paper and no computing equipment.

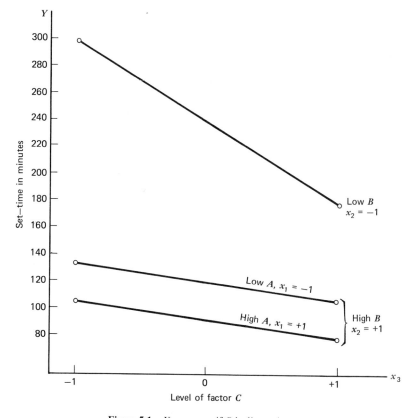

Figure 5.1 Y versus x_3 if B is discontinuous.

The experimenter felt strongly that factor A must have had some effect. He may well have been right. As the reader will note if he carries through the arithmetic work just recommended, the computed A-effect is as large as the C-effect at high B. It does not appear, however, at low B. This state of affairs is shown in Figure 5.1.

But these equations all spring from the premise, contrary to fact, that B is a discontinuous factor. We now return to reality.

5.8. REPRESENTATION WHEN ALL FACTORS ARE CONTINUOUS

When all factors are continuous, the familiar apparatus of quadratic equations is available. But I have rather disingenuously concealed up to this point the fact that a *ninth* observation was indeed taken in this experiment. It was made at the *center* of the design, that is to say, at the midrange of each of the three factors. The response observed was $y_9 = 168$. This is quite close to the average, 172, of the other eight observations.

Suppose that we had tried to fit a general quadratic equation in x_2 and x_3 to the five points in (x_2, x_3) space (since x_1 appears to have a very small coefficient). Thus we would have

(5.4) $$Y = b_0 + b_2 x_2 + b_3 x_3 + b_{23} x_2 x_3 + b_{22} x_2^2 + b_{33} x_3^2.$$

We would have found, almost instantly, that not all of the *six* coefficients can be estimated. We can estimate the first four, however, without even using the fifth data point. We can estimate $(b_{22} + b_{33})$ by $(y_9 - \bar{y})$, where $\bar{y}$ is the average of the first eight y-values. This estimator will be $(168 - 172)$ or -4 in this case. Its standard error will be $(1 + \frac{1}{8})^{1/2}\sigma = 1.06\sigma$ or about 12.7.

Only in an introductory example of the sort under discussion would we dare to "conclude" that this quantity is 0 just because it is about a third of its standard error. A more serious statement about the true value of the sum of the two pure quadratic coefficients would be that it lies, with 95% confidence, between $(-4 + 25.4) = +21$ and $(-4 - 25.4) = -29$. But here, to make other points, we are moving even further in the direction of simplification, and assuming that, since the sum appears to be small, both its summands are 0. We do this in order to show a tolerably simple *response surface*. The reader, however, should not ever make so irresponsible a set of judgments in a real-life case. Do as I say, not as I do.

We are accepting, then, the equation:

(5.5) $$Y = 172 - 66x_2 - 37x_3 + 24x_2 x_3.$$

This equation looks exactly like (5.1), but it has a different meaning. Here x_2 and x_3 are continuous variables, and the equation is that of a continuous response surface.

Figure 5.2 shows a few of its contours. These can be sketched quite easily by solving the equation for x_2, giving Y a fixed value, and then running through a sequence of x_3, x_2 points. Thus:

$$x_2 = \frac{172 - 37x_3 - Y}{66 - 24x_3}.$$

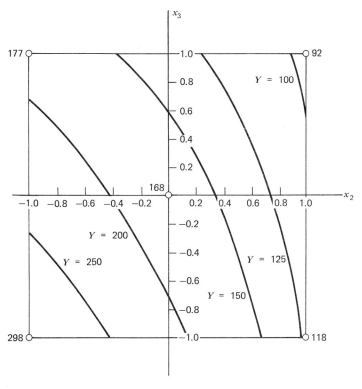

Figure 5.2 Contours of Y when x_2 and x_3 are continuous. $Y = 172 - 66x_2 - 37x_3 + 24x_2x_3$ Circled dots represent data points.

When values for x_2 thus found become too far from the region of interest, it is sensible to use instead the equation for x_3:

$$x_3 = \frac{172 - 66x_2 - Y}{37 - 24x_2}.$$

If less than an hour is required to obtain computing services, any of the many available contour-plotting routines may be used.

5.9.   CONTOURS OF STANDARD ERROR OF Y

The fact that the lines in Figure 5.2 are sharp does not mean that the true contours of constant Y are exactly known. On the contrary, contours of constant $s(Y)$ to be used with Figure 5.2 are given in Figure 5.3. Since the four coefficients in (5.5) all have the same variance, σ^2/N, and are uncorrelated (orthogonal), the equation for the variance of a fitted value, Y, is

$$(5.6) \qquad \mathrm{Var}\,(Y) = \frac{\sigma^2}{N}\,(1 + x_2^2 + x_3^2 + x_2^2 x_3^2) = \frac{\sigma^2}{N}\,\xi^2,$$

when all x_i are coded ± 1 at the data points, and ξ is a multiplier of $\sigma/\sqrt{N}$, also shown in the figure, which can be used in other 2^3's. See Appendix 5.A.

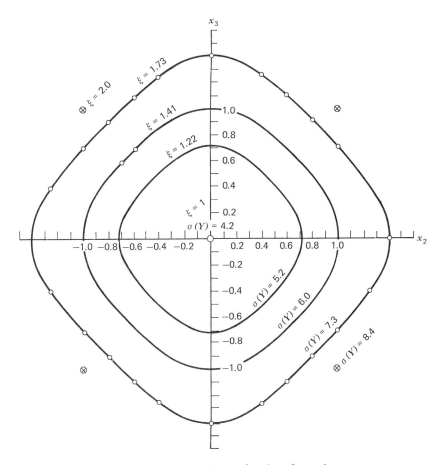

Figure 5.3 $s(Y) = \xi\sigma_y/N$ as a function of x_2 and x_3.

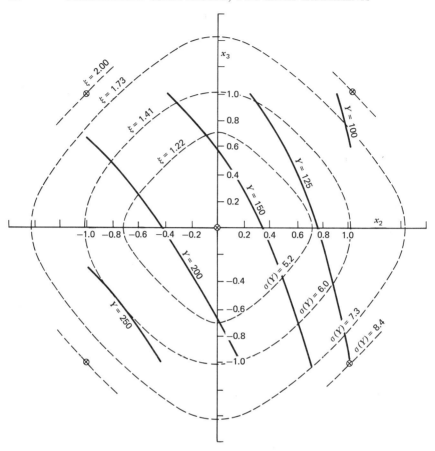

Figure 5.4 Y- and $s(Y)$-contours.

A special case of this equation, for use only at the data points of any $2^p = N$, is:

$$(5.7) \qquad \text{Var} \, (Y \text{ at design points}) = \frac{k\sigma^{2} \, *}{N},$$

where k is the number of constants, including b_0 or $\bar{y}$.

It is of some interest to note that inside the design area the standard error of Y is largest at the data points and is minimal at the center of the design. Some readers will find it easier to look at Figure 5.4, in which the Y-

_____

* I am indebted to Professor Allan Birnbaum for this remarkably useful equation.

contours of Figure 5.2 have been overlaid in light dotted lines with the $\sigma(Y)$ contours of Figure 5.3.

5.10. A NUMERICAL CHECK FOR YATES'S 2^p-ALGORITHM

Here is a numerical check for the 2^3 computation which is simpler but slightly less thorough than that given by Yates:

1. Sum the observations in column 0, in sequence, writing down the subsums under column 0.

   S_8 is the last observation, abc;

   S_4 is $abc + ab$, that is, The sum of every *fourth* observation;

   S_2 is $abc + ab + ac + a$, that is, The sum of every *second* observation;

   S_1 is the sum of *all* observations.

2. After column 1 is completed, sum its upper half. The result should agree with S_1. Then continue summing to get the whole column sum. This must equal $2S_2$.

3. After column 2 is completed, sum its entries from the top. The sum of the first two must be S_1; the sum of the first four must be $2S_2$. The sum of all eight must be $4S_4$.

4. After column 3 is completed, sum its entries, noting the following as you go:

   a. The first entry is S_1.

   b. The sum of the first two is $2S_2$.

   c. The sum of the first four is $4S_4$.

   d. The sum of all eight is $8S_8$.

This check will always catch a single error, but of course, compensating errors are possible. The extension to larger powers of 2 is obvious.

5.11. INTERPRETATION OF THE 2^3

To admit the simple truth, the main interests of the experimenter in these nine data points were (*a*) to see that his equipment was working, and (*b*) to note that the responses were acceptably high. He could then go ahead with his program, which required tests on several types of cement.

The experimenter knew that the form of equation being used could not be exactly right since it implied that Y, the hardening time, would increase at much lower values of x_2 and x_3.

If we compare the observed y-values with those fitted, assuming that B, C, and BC are the only real effects (see Table 5.5), we see that a tolerably good fit has been obtained. Indeed the MS for lack of fit [computed most easily from the differences of the observations in the four cells as $(3^2 + 25^2 + 1^2 + 33^2)/2 \times 4]$ is 216 with four d.f. This is not greatly larger than the expected σ^2 of 12^2 or 144.

If we add the two terms, A and AB, that are next largest, the residual MS is reduced to 8.5. The latter is found from the two contrasts not used in the new equation, namely, (AC) and (ABC), in column 3 of Table 5.3. Thus $(6^2 + 10^2)/8 \times 2 = 8.5$. This smacks of overfitting, but of course nothing can be proved in such a small experiment since we have left ourselves only two d.f. for judging lack of fit.

The reader should take it as an exercise to fit the six-term equation in A, B, AB, C, BC, and the mean, by using the reverse Yates device. He should then compare the fitted values with the observed ones and see whether he likes the new equation better. He will note that the fit is rather strained with only B, C, and BC, since the values at high B apparently fit worse than those at low B. On the other hand, the equation with six effects gives almost too good a fit.

The ambiguity haunting the detailed interpretation of these data is entirely typical of "small" factorials. Quite a large proportion of the effects we are finding are in the doubtful range; both $\hat{A}$ and $\widehat{AB}$ are larger than their expected standard error, but smaller than twice that value. The set of data is not large enough to establish a clear pattern. The detection of revealing patterns, both of concurrence and of discordance, will be one of our main themes when larger sets of data are discussed.

5.12. ONE BAD VALUE IN 2^{3+0}

The artificial data given in column 0 of Table 5.6, panel a, would produce the "effects" shown in the adjacent column (3). On casual inspection it appears that the factors A, B, and C are all influential and additive in their effects. On more thoughtful inspection it appears that all four interaction terms, though small compared to the main effects, have the same magnitude. Only an extremely casual data analyst would pool the four 50's and call the result an unbiased estimate of the random error.

In column 4 the signs of these four are isolated. A gross error in a single piece of data would enter each contrast-sum and influence it by the same absolute amount. Inspection of the rows of the table of signs, Table 5.2, shows that only eight patterns of disturbance (with their negatives) are possible, and that only one of these would produce the pattern of signs of column 4, Table 5.6. We complete the pattern in column 5. If Table 5.2 were not available, we could use the reverse Yates operation on this set of signed ones, as in panel b of Table 5.6, to find that only run a could be responsible, and that it must be off by $+50$. We can revise the column of effects

TABLE 5.6.
PATTERNS IN THE EFFECT CONTRASTS JUDGED NONSIGNIFICANT

Panel a								Panel b					
Spec.	(0)	(3)	(4)	(5)	(6)	(6) ÷ 8				Reverse Yates			
(1)	158	1650		+	1600	200		ABC	+1	2	0	0	abc
a	132	−254		+	−304	−38	A	BC	+1	−2	0	0	bc
b	212	166		−	216	27	B	AC	−1	−2	0	0	ac
ab	136	−50	−	−	0	0		C	−1	2	0	0	c
c	264	374		−	424	53	C	AB	−1	0	−4	0	ab
ac	188	−50	−	−	0	0		B	−1	0	4	0	b
bc	318	50	+	+	0	0		A	+1	0	0	8	a
abc	242	50	+	+	0	0		T	+1	0	0	0	(1)

directly, without repeating the whole computation, by adding 50 to or subtracting 50 from each of its members with signs reversed from those in column 5. In this way we get the revised contrast-sums shown in column 6, panel a.

Our conclusion about the magnitude of the random error is now entirely different. Each of the effects has been increased. Of course with real data the nonrandomness induced by a bad value will not be so obvious since it will be somewhat obscured by random error.

A heavy price must be paid for this piece of cleverness, just as it would be if one value were missing altogether. Although the computation of Table 5.6 gives no hint of the fact, all effects are now estimated with *twice* the variance of the unmodified set. P. W. M. John [1973] was the first to point out this rather horrifying fact, but he also shows some ways of ameliorating the loss in the larger factorials.

Since we cannot hope to estimate all eight factorial parameters from seven observations, the natural one to forgo is ABC. This is equivalent to estimating each of the other effects plus or minus ABC. We can tease each of these out by reverse Yates, asking, for example, what combination of observations estimates $A - ABC$. A simple rule emerges: A different set of four observations is required for each estimate; those for main effects are in a pair of diagonally opposite edges that do not include the bad value, and those for two-factor interactions are in the face that varies the two components but is away from the bad value. Thus in Figure 5.5a, for factor A and for observation (1) to be excluded, we use the four points marked in the "diagonal square": ab, b, ac, c. For the 2fi AB the four points marked in Figure 5.5b should be chosen. It should be obvious or verifiable that the contrast $(abc - bc) - (ac - c)$ has expected value $4(AB + ABC)$.

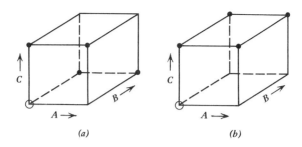

Figure 5.5   Data points for estimation of (*a*) *A* and (*b*) *AB* in a $2^3 - 1$.

5.13.   BLOCKING THE 2^3

The familiar advantage of blocking (and often the only one mentioned) is the possibly improved precision if the blocking isolates sets of experimental material that are more homogeneous than the whole. There will be many cases in which two litters of four are easier to obtain than one of eight, and many in which two smaller batches of raw material will give better comparisons within batches than will one large batch, or a random assignment of runs to material regardless of batch. But the other side of blocking needs more emphasis.

Even when all eight runs could be run on one batch of raw material, there is an important reason for *not* doing so. We are always trying to get conclusions that are widely applicable. If we confine all the data to a single batch, and then suppose that our conclusions must hold for all batches we make a needlessly reckless extrapolation. Our results will be as precise, and our base broader, if we plan to take half the data on one batch and half on a batch separated as widely as possible from the first.

Of course something must be sacrificed if we are to eliminate batch-to-batch differences from our comparisons. It is usually thought safest to confound the *ABC* interaction with batch differences by doing in one batch the runs that are on the "plus" side of the *ABC* contrast. This requires doing (1), *ab*, *ac*, and *bc* in one batch, and *a*, *b*, *c*, and *abc* in the other. These are sometimes called the "even" and "odd" blocks for the number of letters in the specification of each run.

A conscientious seeker after truth will remember that the blocks should be done in random order, and that the set of four runs inside each block should also be objectively and separately randomized. The tables of Moses and Oakford [1963] provide a very convenient way to do this.

Unfortunately there are not many choices for such small sets, and many of them are distasteful because they are sensitive to any linear or simply curved trend in time. Some methods for avoiding these difficulties will be discussed in Chapter 15 on trend-free plans.

When blocks of *two* experimental units are called for, an unreplicated 2^3 cannot be expected to give much information on interactions. The interaction contrasts will generally be aliased with block differences, and so the first set of four blocks will measure only main effects with within-block precision, while the two-factor interactions will be measured with the *among*-block variance. The four blocks, each a pair of points at the ends of a different major diagonal of the cube, are:

I. (1), *abc*

II. *a, bc*

III. *b, ac*

IV. *c, ab*

Estimates of the three main effects, freed of any aliasing with 2fi's but not of the 3fi *ABC*, can be made with two-thirds efficiency after any *three* of these blocks have been completed. Other sets of blocks must be done if the three 2fi's are to be estimated with the within-block variance. See Sections 10.3.2 and 10.3.3.

5.14. SUMMARY

Much of the content of this chapter is standard and is a reflection of the treatment in many earlier works. Much of it is clearly presented in Yates's pamphlet [1937]. Davies [1971] contains a good exposition of the 2^3, although the factorial representation (displayed on page 235 of that work) is not employed in the body of the text, nor are data used there to derive fitting equations.

The representation of the response surface as a factorial function of the factor levels is not new, but is carried through here in elementary detail for several cases (Sections 5.7 and 5.8) for experimentalists and engineers who may not have seen the statistical textbooks or papers in which this is explained.

The numerical check for Yates's algorithm is new but owes something to R. Freund, who showed me a similar one long ago. The search for "bad values" in Section 5.12 has not, I believe, been published elsewhere. The emphasis on blocking for generality rather than for precision is new, or at least original.

But, as I warned in Section 5.1, the 2^3 is not widely recommended. It is treated here at such length as a miniature exemplification of the 2^p plans, to introduce several sets of nomenclature and several modes of representation of the facts that are unfamiliar to most experimenters. Every one of the topics introduced in the sections of this chapter will appear repeatedly in later chapters.

APPENDIX 5.A

THE VARIANCE OF LINEAR FUNCTIONS OF UNCORRELATED RANDOM VARIABLES

Most readers already know the results of this appendix. Those who do not, however, must acquire them to profit fully from this chapter and from most of those to follow.

I assume first that y is a random variable with expectation μ and variance σ^2. The variance of a random variable is defined as the expected value of its *squared deviation from its expected value*. Thus:

$$E\{y\} \equiv \mu, \qquad E\{(y - \mu)^2\} \equiv \sigma_y^2.$$

We take the simple linear function of y, $z = a + cy$, where a and c are known constants, and we write in turn the expected values of z and of its squared deviation from its expectation:

$$
\begin{aligned}
E\{z\} &= E\{(a + cy)\} \\
&= E\{a\} + E\{cy\} && \text{since } E\{\ \} \text{ is a linear operator} \\
&= E\{a\} + cE\{y\} && \text{since } E\{\ \} \text{ is a linear operator} \\
&= a + c\mu;
\end{aligned}
$$

$$
\begin{aligned}
\text{Var}(z) &= E\{(z - E\{z\})^2\} \\
&= E\{(a + cy - a - c\mu)^2\} \\
&= E\{[c(y - \mu)]^2\} \\
&= c^2 E\{(y - \mu)^2\} && \text{since } E\{\ \} \text{ is a linear operator} \\
&= c^2 \sigma_y^2.
\end{aligned}
$$

In summary, we may write

(5.A.1) $$\text{Var}(a + cy) = c^2 \sigma_y^2.$$

We extend this to find the variance of a weighted sum of n uncorrelated random variables y_i, which may have different expectations, different variances σ_i^2, and different weights c_i.

(5.A.2) $$\text{Var} \sum_{i=1}^{n} c_i y_i = \sum_{1}^{n} c_i^2 \sigma_i^2.$$

If the weights are all set at 1,

$$(5.A.3) \qquad \text{Var} \sum_1^n y_i = \sum \sigma_i^2.$$

In words, the variance of a sum of uncorrelated random variables is the sum of their variances.

If each weight is set at $1/n$, and if $\sigma_i^2 = \sigma_y^2$ (constant),

$$\text{Var} \sum c_i y_i = \text{Var} \frac{1}{n} \sum y_i = \text{Var} \bar{y} = \frac{1}{n^2} \sum \sigma_i^2 = \frac{\sigma_y^2}{n}.$$

The equation

$$(5.A.4) \qquad \text{Var} \bar{y} = \frac{\sigma_y^2}{n}$$

for uncorrelated or independent y, all with the same variance, is perhaps the most important equation in statistics.

Perhaps the second most useful equation, at least in our work, is

$$(5.A.5) \qquad \text{Var} \sum c_i y_i = \sigma_y^2 \sum c_i^2,$$

which again holds when all y_i are uncorrelated with the same variance but not necessarily with the same population means.

Applying (5.A.5) to a general equation like (5.5), i.e. to:

$$Y = \bar{y} + b_1 x_1 + b_2 x_2 + b_{12} x_1 x_2;$$

and, remembering that the b's and $\bar{y}$ are the random variables, all uncorrelated (orthogonal) and all of the same variance, σ^2/N, while the x's are exact, we have

$$\text{Var } Y = \text{Var } \bar{y} + x_1^2 \text{ Var } b_1 + x_2^2 \text{ Var } b_2 + x_1^2 x_2^2 \text{ Var } b_{12}$$
$$= \frac{\sigma^2}{N} (1 + x_1^2 + x_2^2 + x_1^2 x_2^2),$$

which is (5.6).

CHAPTER 6

Unreplicated Four-Factor, Two-Level Factorial Experiments

6.1.   INTRODUCTION

Sixteen-run plans are generally much more valuable than eight-run plans. The gain does not occur primarily because of the improved precision of the larger plans, nor is it heavily influenced by the larger number of independent contrasts available for estimating effects and interactions. I believe that it is due mainly to the opportunity for *studying* the data—to learn about the presence of one or more defective values, about possible transformations, and about the randomization pattern and its defects—which just begins to appear in the 2^4.

6.2.   THE FIRST COMPUTATIONS

The data come from a well-executed unreplicated four-factor experiment. They were the first observations taken on a prototype piece of equipment for which no error estimate was available. The experimenters hoped for a standard deviation of about 5%. Three of the factors were known to have positive effects: A, the load on a small stone drill; B, the flow rate through it; and C, its rotational speed. The fourth factor, D, was the "type of mud used in drilling."

The response was the rate of advance of the drill. Table 6.1 gives the data in column 0, and shows the details of the usual computation to arrive at the contrast-sums listed in column (4).

TABLE 6.1.
STANDARD FIRST STAGES IN THE NUMERICAL ANALYSIS OF A 2^4

Spec.	Obs. (0)	(1)	(2)	(3)	(4)	Effect
(1)	1.68	3.66	10.38	40.10	98.48	Total
a	1.98	6.72	29.72	58.38	7.30	A
b	3.28	10.68	13.13	0.28	26.38	B
ab	3.44	19.04	45.25	7.02	1.20	AB
c	4.98	4.51	0.46	11.42	51.46	C
ac	5.70	8.62	−0.18	14.96	4.76	AC
bc	9.97	17.20	0.81	−1.76	12.04	BC
abc	9.07	28.05	6.21	2.96	1.34	ABC
d	2.07	0.30	3.06	19.34	18.28	D
ad	2.44	0.16	8.36	32.12	6.74	AD
bd	4.09	0.72	4.11	−0.64	3.54	BD
abd	4.53	−0.90	10.85	5.40	4.72	ABD
cd	7.77	0.37	−0.14	5.30	12.78	CD
acd	9.43	0.44	−1.62	6.74	6.04	ACD
bcd	11.75	1.66	0.07	−1.48	1.44	BCD
abcd	16.30	4.55	2.89	2.82	4.30	ABCD

There are many ways of judging the set of 15 factorial effects, none of them objective. The commonest method is to take an oath before the data are taken that we will use the three- and four-factor interactions as estimates of error, and to then use "F- or t-tests" on each of the remaining contrasts, judging all those to be significant which are larger than a critical F-value, usually taken as the 5% point for the number of degrees of freedom pooled for error. This method, recommended in most textbooks, is frequently violated as soon as the data are in, first of all by the use of several levels of significance to indicate which effects are more and which less "significant." The lack of seriousness of the whole enterprise is revealed by the fact that no statistician has thought to investigate the operating characteristic (frequency of missing real effects) of the combined multilevel test. Examples will be given later of entirely jejune conclusions drawn in this way.

At this point only a commonsense inspection of the relative magnitudes of the contrasts is urged. It is entirely obvious that the five largest effects—B, C, BC, D, and CD—are dominant as well as plausible. They are dominant because of the wide gap (of 4.74) separating them from the remaining 10 contrasts, which are all smaller than 7.32. They are plausible because the two-factor interactions contrasts—BC and CD—are nonadditivities of pairs of the most influential factors.

One way of expressing the dominance of the five largest contrasts is by computing the "coefficient of determination," R_y^2. This is the fraction of the total scatter of the original 16 observations, expressed as a sum of squares about their mean, that is accounted for by the five effects we have chosen. The total sum of squares (abbreviated as TSS henceforth) of deviations of our observations about their mean is 262.68. The SS accounted for by the five effects chosen is $(26.38^2 + 51.46^2 + \cdots)/16$ or 249.15. Then R_y^2 is 249.15/262.68 or 0.948.

We seem to have done quite well, but of course we need some control since we have chosen the five largest effects after the fact. A rough computation (using Owen's tables [1962] to find expected values and variances of normal order statistics) indicates that the five largest of 15 normally distributed values may be expected to account for 75% of their TSS.

If we assume—as we are not really entitled to—that the residual SS, 13.52, measures random error, then 13.53/10 or 1.35 is an estimate of the error variance. Since the mean square for the five effects is 249.15/5 or 49.83, we see that the ratio of these two MS's, one for effects and one for random error, is 49.83/1.35 or 36.8. There is little point in judging the "statistical significance" of this value by reference to F-tables. In the first place, we may have *under* estimated the error variance by our arbitrary assignment of BC to the set of real effects. In the second place, we may have *over* estimated the error variance by leaving some real effects in the residual SS. Third, we compute this value only to compare it with later ratios of the same sort. It is not our aim to develop a test for the significance of the difference in significance of two significance tests.

Table 6.2 shows in detail (for the last time) the "reverse Yates computation" to get fitted values Y. The residuals d_y are calculated in the last column of the same table. Some readers may find it interesting to compute the residuals directly, without the computation of the Y, by dropping the effects judged real from the column of effects, and carrying the reverse Yates operation through only on the effects judged *not* real. The values in the last column of this operation must of course be divided by N, the number of observations, to give the residuals in original units.

We now explore in some detail the consequences of judging that factors B, C, D are the only influential ones, and that B, C, BC, D, and CD are the

TABLE 6.2.
"Reverse Yates" Computation of Fitted Values Y and of Residuals d_y

Effect	Contrast-Sum	(1)	(2)	(3)	(3) ÷ 16	+6.2 = Y	Spec.	y	d_y
BCD	0	13	31	120	+7.5	13.7	bcd	11.8, 16.3	−1.9, +2.6
CD	13	18	89	44	+2.8	9.0	cd	7.8, 9.4	−1.2, +0.4
BD	0	63	31	−32	−2.0	4.2	bd	4.1, 4.5	−0.1, +0.3
D	18	26	13	−60	−3.8	2.4	d	2.1, 2.4	−0.3, 0.0
BC	12	13	5	58	+3.6	9.8	bc	10.0, 9.1	+0.2, −0.7
C	51	18	−37	−18	−1.1	5.1	c	5.0, 5.7	−0.1, +0.6
B	26	39	5	−42	−2.6	3.6	b	3.3, 3.4	−0.3, −0.2
T	0	−26	−65	−70	−4.4	1.8	(1)	1.7, 2.0	−0.1, +0.2

only real effects. This is done by study of the empirical cumulative distribution (abbreviated as e.c.d. from now on) of the residuals and of their possible relations with their corresponding Y-values. Ideally, the the e.c.d. should approximate a normal one, and there should be *no* visible relation of the d_y to the Y.

Figure 6.1 shows the 16 residuals from $Y(\bar{y}, \hat{B}, \hat{C}, \widehat{BC}, \hat{D}, \widehat{CD})$ plotted as an e.c.d. on a "16-residual normal grid."* The d_y are first entered directly near the left margin of the grid. They are then easily moved in, each one going to the succeeding line on the grid. Figure 6.2 is an "Anscombe-Tukey plot" [1963] of all 16 residuals against their corresponding Y-values. Both plots fail to conform to our hopes. The e.c.d. does not give a straight line; there is a clear trend to increasing d_y with increasing Y. We will discuss these matters somewhat more objectively in the next section, after reviewing here the numerical operations that should become routine.

The work may be listed under five steps:

1. Yates's algorithm on y to get effects.
2. Reverse Yates on large effects to get Y.
3. Computation of d_y.
4. Normal plot of d_y.
5. Plot of d_y versus Y.

The choice of terms to put into the fitting equation was subjective but rather easy in this example. The examination can be made more thorough when fast computing service is available. First the forward computation of

---

* Two full-size blank grids are included for copying.

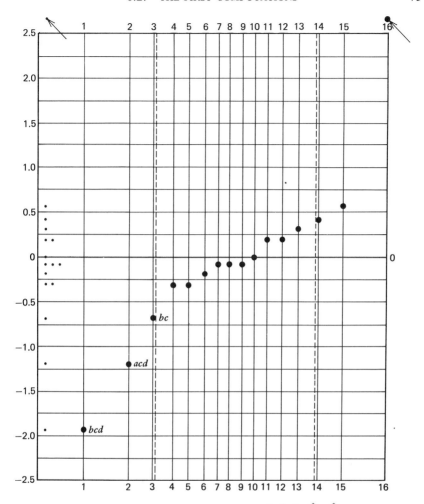

Figure 6.1 Residuals from equation $Y(\bar{y}, \hat{B}, \hat{C}, \hat{D}, \widehat{BC}, \widehat{CD})$.

the whole set of effects is carried out. Then a sequence of fitting equations is run through, each equation adding the next smaller effect, regardless of its meaning, to the fitting equation. If this is carried through automatically, without human intervention or arbitrary stopping rule, for, say, *half* of the total number of effects, a lot of paper will be wasted, but 99% of all the experiments I have seen will have been covered. As very crude guesses, about four real effects is average for a 2^4, and seven for a 2^5. The e.c.d. and d_y versus Y plot must be made for each fitting equation. These can be (and indeed have been) programmed for many computers.

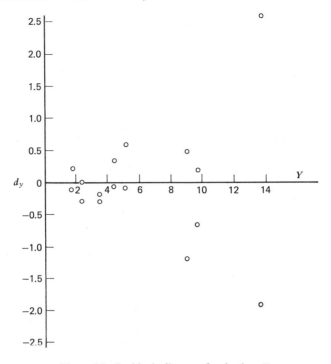

Figure 6.2   Residuals $d\hat{y}$ versus fitted values Y.

If this is done for the present 2^4, it will be found that the largest residuals continue to appear excessive in the e.c.d. plot, and that the dependence of d_y on Y persists.

6.3.   INTERPRETATION OF THE FIRST COMPUTATIONS

6.3.1.   The Empirical Cumulative Distribution of the Residuals

The vertical lines of the "16-residual normal grid" of Figure 6.1 are spaced to give a straight line—on the average—when a set of normally distributed independent values is ordered and plotted. But of course a particular sample of 16 numbers derived from observations never gives an exact straight line. In addition to the inevitable fluctuations of samples of *independent* normal deviates, our sets will always have a further distortion caused by their being residuals from a fitted equation and hence not independent.

I see no prospect for gaining insight into the behavior of sets of residuals by study of the variances and covariances of the usual normal order statistics as they are published. There is at present no substitute for the direct examination of a considerable collection of e.c.d.'s of the size we need. We would like a set originating from independent random normal deviates, and another set from fitted equations. Appendix 6.A gives 40 plots of 16 independent standard normal deviates (r.s.n.d.) as drawn from the Rand tables [1955, pages 1–200] with nothing removed. If five or so d.f. are removed from a set of r.s.n.d. by multiple regression, some of the irregularities of the original e.c.d. will be removed, so that the normal slot will appear more nearly normal, even "supernormal," that is, with extreme deviates smaller than expected. Although it is salutary to note the variations in shape that these little sets show, none has the shape of Figure 6.1.

I conclude that something is amiss. *Either* there are two excessive values which were disturbed by some factors that did not operate on the other 14 observations, *or* the random error is not normally distributed, *or* the factorial representation is far from the best, *or* the response is not in the right units. We must strongly resist accepting the first alternative. It should be considered only if all else fails.

6.3.2. The d_y versus Y Plot

Figure 6.2 shows heavy dependence of the magnitude of d_y on Y. This dependence is of the simplest kind, suggesting that d_y increases with the true value of y. The true value of y is often designated as η (read "eta"). If the "per cent error," or the "coefficient of variation" σ/η, is constant, we expect that log y will have constant error. A constant coefficient of variation (usually abbreviated as CV) implies that the uncontrolled factors producing random variation somehow know how large the quantity is that they are disturbing. The simplest assumption needed to account for this behavior is that the random factors are operating multiplicatively, rather than additively. If the same mode of influence extends to the controlled factors, we may hope that some or all of the interactions found in fitting an additive equation, as we did above, will be removed.

Using log y as a response, after a study of y shows residuals increasing with Y, is not simply a lazy statistician's way of reinstating his assumption of constant average error. If its use simplifies the fitting equation, if a notably better fit is obtained, *and* if the dependence of d_y on Y is removed or greatly decreased, we can feel quite sure that log y is a more informative response than y.

Many properties of physical systems meet the obvious mathematical requirements for "logging." The response must of course be intrinsically positive. The ratio of the largest response to the smallest should be fairly large, say at least 5:1. These conditions are met by the present data. Indeed, when speedy computer service is available, my own practice is to ask for the whole sequence of computations and plots listed in the preceding section, *both* for y and for log y.

6.4.   LOOKING FOR SIMPLE MODELS

Table 6.3 and Figures 6.3 and 6.4 show the results of the usual computations, using $z = 100 \log_{10} y$ as a response. We have simplified our equation, and we have improved our fit. Main effects A, B, C, D now give an R_z^2 of 0.986; adding CD raises this to 0.9908. But we still have one suspiciously large residual—that at bcd—and we still have a whole collection of residuals that increase with Z, though perhaps less than they did in the corresponding situation with Y.

I have not been entirely fair to the experimenters in my rather slow approach to "logging." They knew that many factors operated exponentially on the rate of advance y, but they did not think that as small a set of observations as this one could throw much additional light on the matter. Two suggestions emerged after discussion: It is possible that some low rate of drilling is itself a zero level, and it is conceivable that some fixed low power of y is the response on which the factors might operate additively.

TABLE 6.3.
EFFECTS ON $z = 100 \log_{10} y$; FITTED VALUES $Z(\bar{z}, A, B, C, D, CD)$;
RESIDUALS d_z

Panel a Spec.	y	z	(4)	Effects	Panel b (0)	Z	d_z
(1)	1.68	22.5	1110.0	T	1110	23.9	-1.4
a	1.98	29.6	45.2	A	45	29.6	0.0
b	3.28	51.6	201.8	B	202	49.2	$+2.4$
ab	3.44	53.7	-11.8	AB	0	54.8	-1.1
c	4.98	69.7	401.0	C	401	69.8	-0.1
ac	5.70	75.6	3.8	AC	0	75.5	$+0.1$
bc	9.97	99.8	-17.6	BC	0	95.1	$+4.7$
abc	9.07	95.8	3.6	ABC	0	100.7	-4.9
d	2.07	31.6	113.4	D	113	33.8	-2.2
ad	2.44	38.7	23.0	AD	0	39.4	-0.7
bd	4.09	61.2	-5.2	BD	0	59.1	$+2.1$
abd	4.53	65.6	18.0	ABD	0	64.7	$+0.9$
cd	7.77	89.0	34.0	CD	34	88.2	$+0.8$
acd	9.43	97.4	18.4	ACD	0	93.8	$+3.6$
bcd	11.75	107.0	-11.8	BCD	0	113.5	-6.5
$abcd$	16.30	121.2	13.4	$ABCD$	0	119.1	$+2.1$

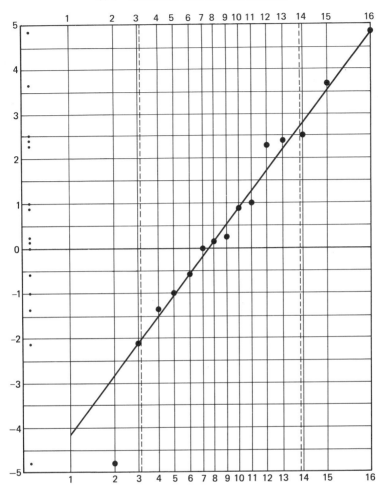

Figure 6.3 Residuals from equation $Z(\bar{z}, \hat{A}, \hat{B}, \hat{C}, \hat{D}, \widehat{CD})$; $100 \log_{10} y = z$.

The first possibility can be tested by trying $\log(y - c)$ as a response, letting c take each of several values, say 0.5, 1.0, and 1.25. Figures 6.5 and 6.6 show the residual e.c.d. and d_w versus W plots for $w = 100 \log 100(y - 1)$. The second set of alternatives can be canvassed by using y, $y^{1/2}$, $\log y$, $y^{-1/2}$, and y^{-1} as responses. Table 6.4 gives a summary of all nine cases. In the first five we have varied the exponent of y; in the latter five (case VII $\equiv$ case III) we have tried five values of c.

We see that $\log y$ does best if we look only at overall fit to four main effects. But $y^{-1/2}$ does somewhat better if one two-factor interaction is allowed. *It*

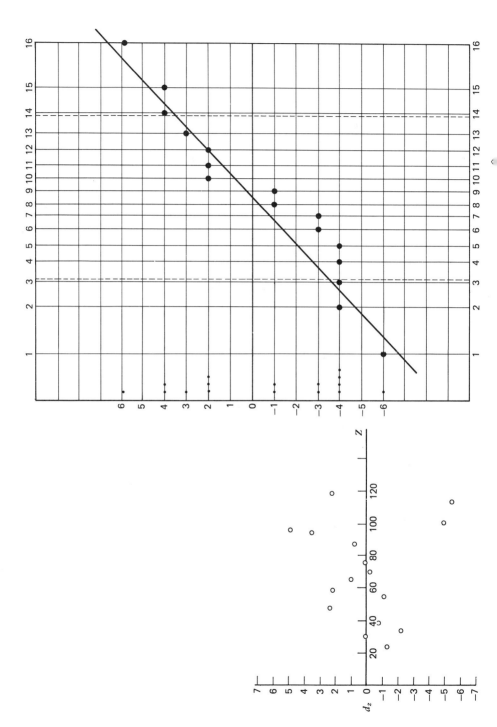

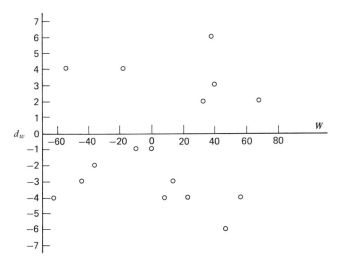

Figure 6.6 Residuals d_w versus fitted values W;

TABLE 6.4.
SUMMARY OF A 2^4 UNDER NINE y-TRANSFORMATIONS

Transform	Response	$R_1^2(A, B, C, D)$	Bad Values	d_y vs. Y	CV	Added Interactions	R_2^2
I	y	0.8878	bcd, abcd		19%	BC, CD	0.9485
II	$y^{1/2}$	0.9562	abcd			CD	0.9757
III	$\log y$	0.9856*	bcd			CD	0.9908
IV	$y^{-1/2}$	0.9683	†	OK		BC	0.9930*
V	y^{-1}	0.9168	†			BC	0.9857
VI	$\log(y + 1)$	0.9828				BC	0.9902
VII (=III)	$\log y$	0.9856*	bcd			CD	0.9908
VIII	$\log(y - 0.5)$	0.9851				BC	0.9898
IX	$\log(y - 1.0)$	0.9777		OK	11%	BC	0.9915
X	$\log(7 - 1.25)$	0.9673				BC	0.9891

* Highest R^2 in its column.
† Obviously nonnormal because of absence of BC.

has the further advantage that the residuals from this equation show no trend with response. There is little to choose between $y^{-1/2}$ and $\log(y - 1)$, so far as our analysis shows. Both require *BC*, and neither shows any excessive residuals or trend of residuals with response.

This is an appropriate time at which to return the data and the analysis to the experimenters. It is they who will decide whether the equation in $y^{-1/2}$ is intriguing or is merely a curiosity. A summary of our findings can be written entirely in engineering terms, with no more mathematical apparatus than elementary algebra and tables of data and residuals.

6.5. A NOTE ON ROUNDING IN YATES'S ALGORITHM

I am indebted to Ms. E. (Reid) Flaster for an ingenious suggestion that often saves time in hand or desk-machine computation of the 2^p factorial effects. Suppose that we round the y-data of Table 6.1 to the nearest integer and then carry through the usual four-column calculation. The result will be the column label "(4)" in Table 6.5, panel a. We see immediately that there are too many ties. These are two 0's, three 4's, four 6's, and two 12's. These numbers are not behaving like a set of random variables from a continuous distribution.

We can carry through a set of one-digit corrections as in panel b of the same table. The new column 4 values designated as (4), are then (10 times)

TABLE 6.5.

A CRITERION FOR ROUNDING RESPONSES IN A 2^4

Panel a. Rounding to digits produced too many ties
Panel b. Revision to one decimal reduces ties to unimportance

(0)	(1)	(2)	(3)	(4)	(0)'	(1)'	(2)'	(3)'	(4)'	(4) + 0.1(4)'
2	4	10	40	98	−3	−3	4	2	6	98.6
2	6	30	58	6	0	7	−2	4	10	7.0
3	11	13	0	26	3	−3	1	2	4	26.4
3	19	45	6	2	4	1	3	8	−8	1.2
5	4	0	10	52	0	5	4	14	−4	51.6
6	9	0	16	4	−3	−4	−2	−10	8	4.8
10	17	1	−2	12	0	2	−3	2	2	12.2
9	28	5	4	0	1	1	11	−10	14	1.4
2	0	2	20	18	1	3	10	−6	2	18.2
2	0	8	32	6	4	1	4	2	6	6.6
4	1	5	0	6	1	−3	−9	−6	−24	3.6
5	−1	11	4	6	−5	1	−1	14	−12	4.8
8	0	0	6	12	−2	3	−2	−6	8	12.8
9	1	−2	6	4	4	−6	4	8	20	6.0
12	1	1	−2	0	−2	6	−9	6	14	1.4
16	4	3	2	4	3	5	−1	8	2	4.2

the corrections to column 4 of panel *a*. These are added in the last column of the table to give us closer estimates of the correct values. We note that there are now only two pairs of ties, one at 1.4 and one at 4.8.

A set of corrections to 0.01 will surely not change any of these values greatly. Even if all the revisions were of size 0.04, their maximum possible effect on a value in column 4 would be 16×0.04 or 0.64. This is only half the smallest of our panel *b* results.

6.6. SNARES (AND DELUSIONS)

A parting glance at the computation in Table 6.1 may induce the reflection that *all* the contrasts in column 4 should not be positive. Even though three of the four factors were known to have positive effects, it is not credible that all the interactions be positive. Nor is it likely that all the error contrasts, which should have an expected value of 0, will be on one side of 0.

In casting about for the simplest possible cause for this run of plus signs, the reader may come upon the fact that an excessive result for one and only one of the 16 observed responses—that at *abcd*—would bias all the contrasts in the same direction. It is tempting to revise this offending value by an amount that would make the set of 15 contrasts (or perhaps only the 9 smaller inter-action contrasts) look more like a set of random normal deviates. *This should not be done.*

If the pattern of signs in column 4 indicated a bad value anywhere but at *abcd*, a more persuasive case for revision could be made. But just here we have the extreme condition—high levels for all factors, all with positive average effects. By inspection we see that all the other results lead up to this one. It would be too bad to lose the knowledge that we really can drill fastest under these conditions just to enforce a good fit for a rather arbitrary form of equation. If the large discrepancies had appeared at intermediate values of the response, our hesitation in criticizing or even revising them would have been less. In positive terms, an interior point, either in factor space or in response range, is more safely judged to be excessive than is an extreme one. We note, finally, that the predominance of long runs of signs in the effect column is removed by the transformation to $\log y = z$ (Table 6.3).

Sermon III is only a short reiteration of points already made: There is much information in large sets of balanced data, not only about the effects of the factors varied, but also about the appropriateness of the fitting equation and about the form of the error distribution. The balance and symmetry of the 2^p plans give them properties of *response by pattern* to many sorts of dis-turbance. In the example just treated, such patterns showed in the behavior of whole sets of residuals under various transformations of the response. There will be many more.

EMPIRICAL CUMULATIVE DISTRIBUTIONS, EACH OF
16 INDEPENDENT STANDARD NORMAL DEVIATES

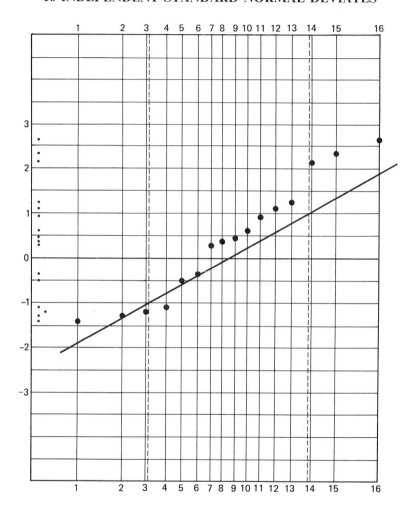

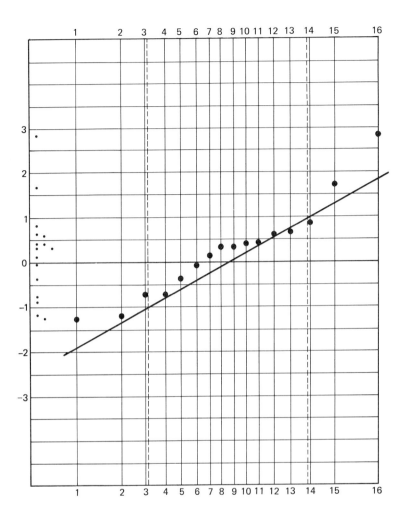

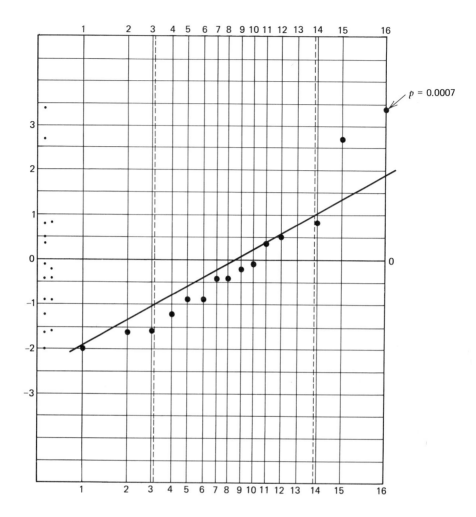

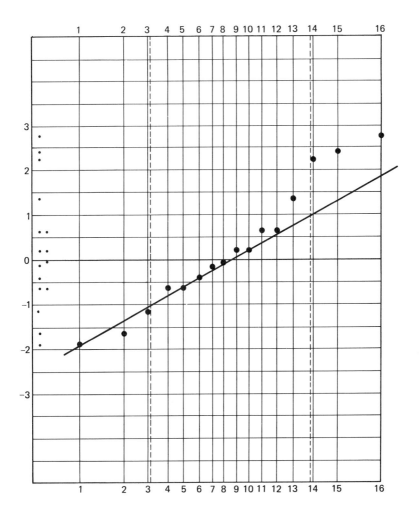

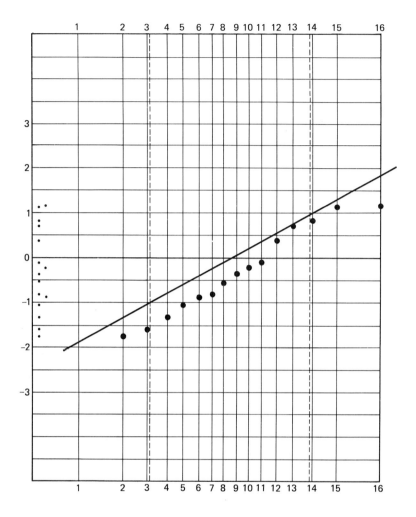

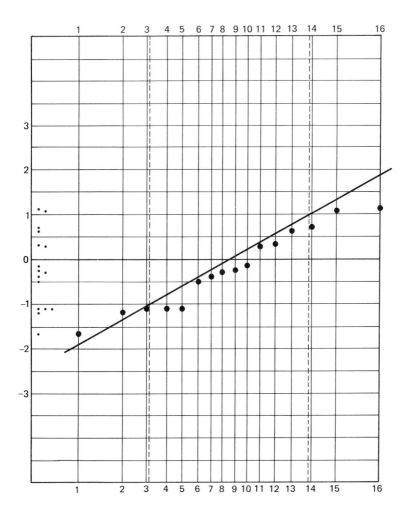

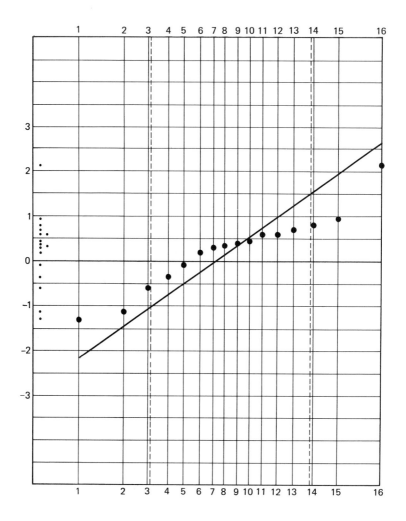

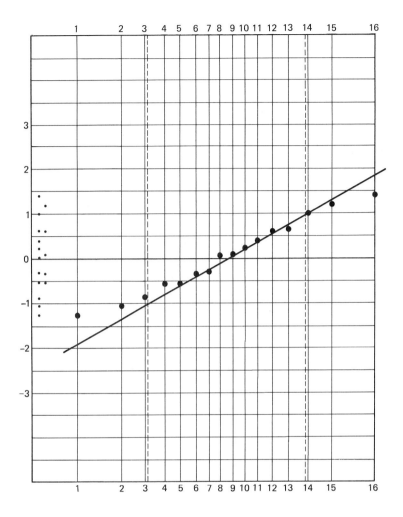

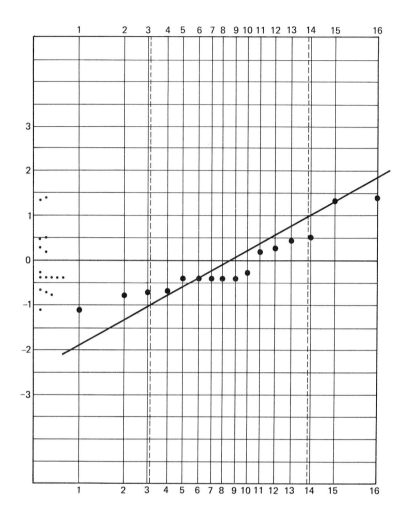

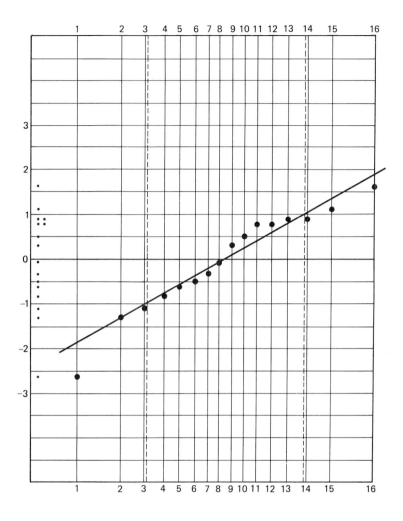

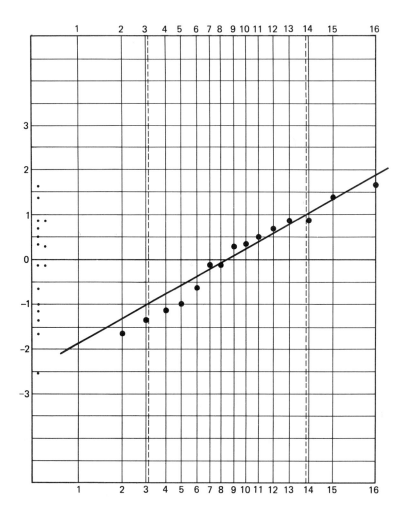

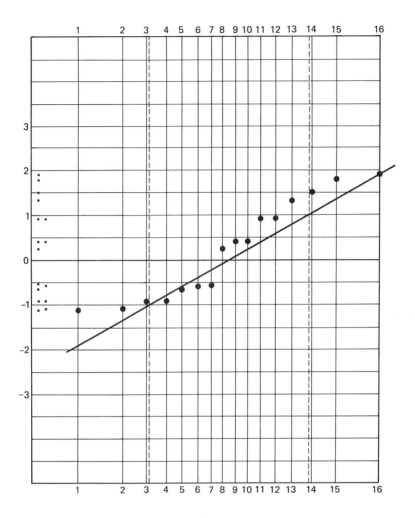

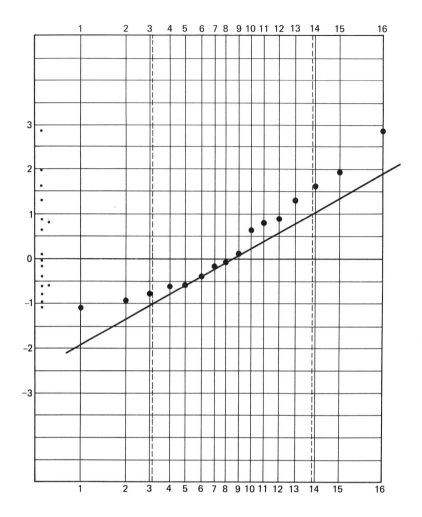

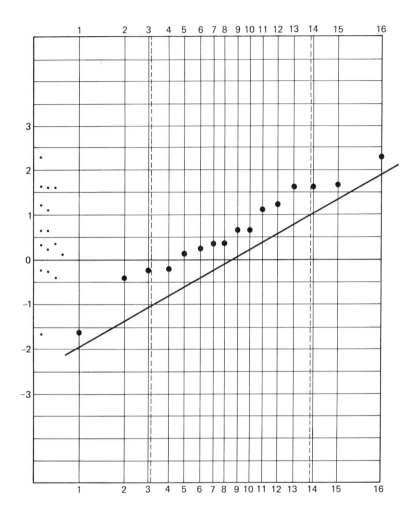

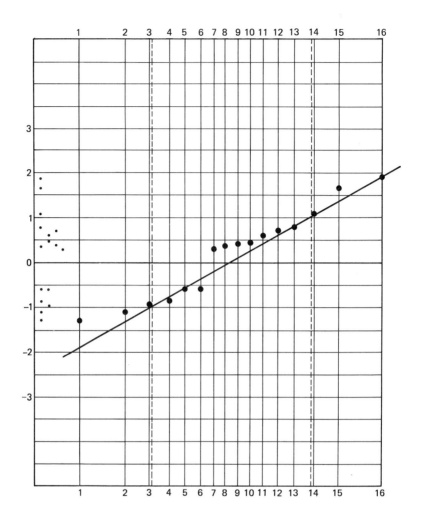

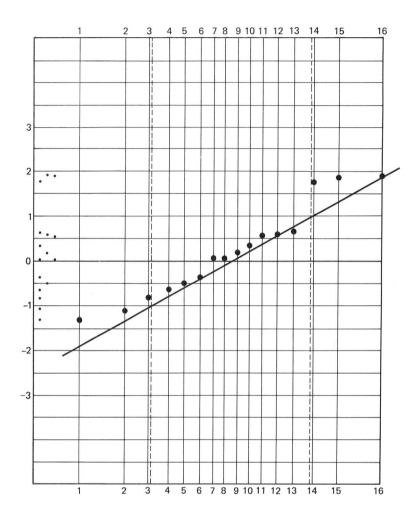

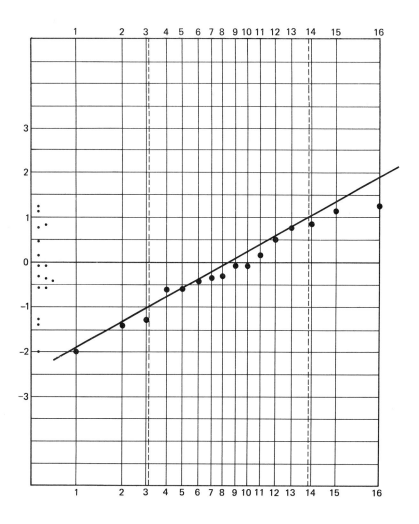

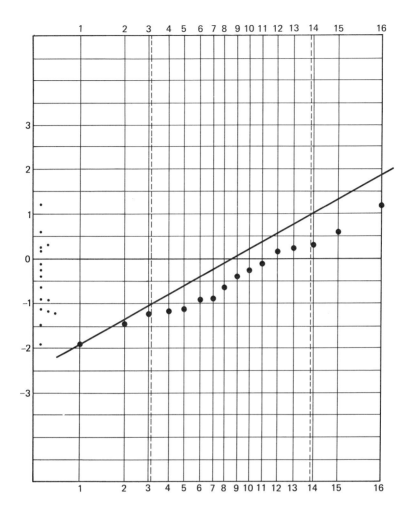

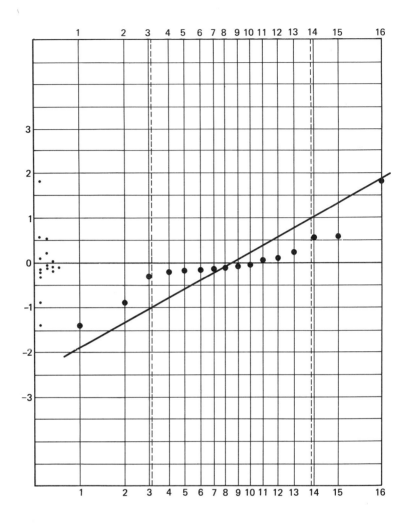

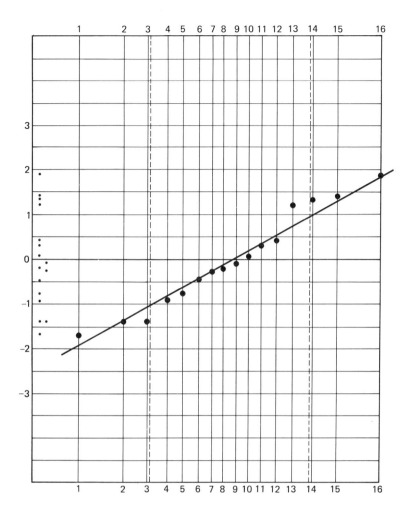

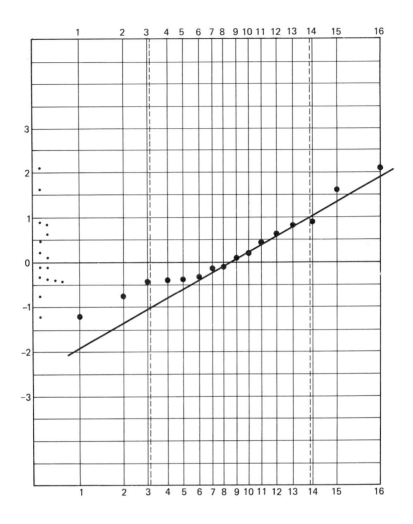

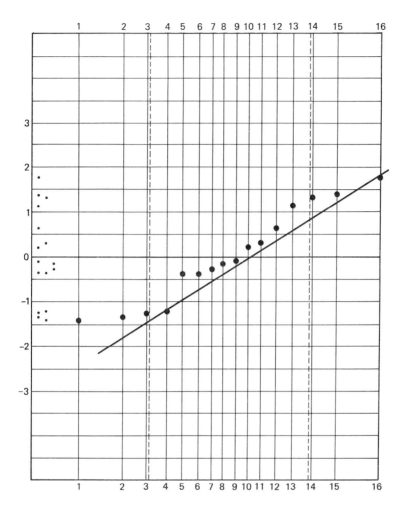

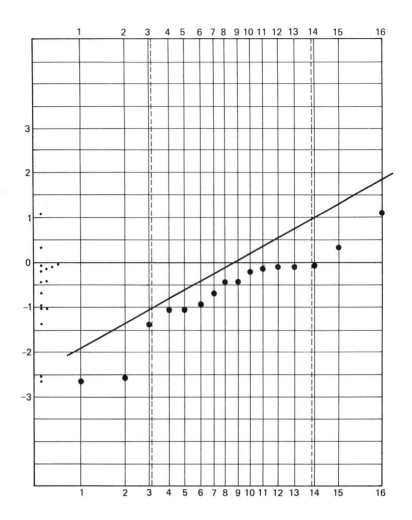

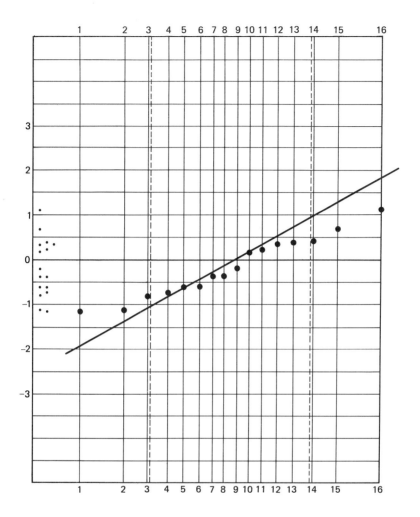

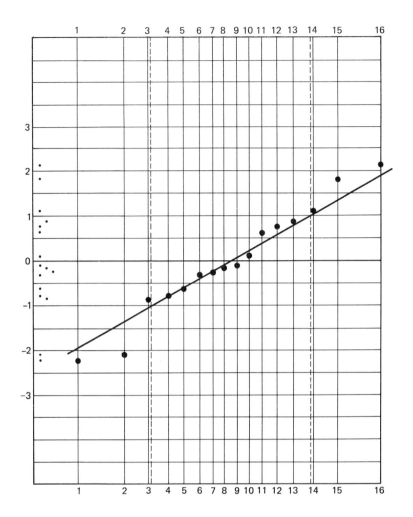

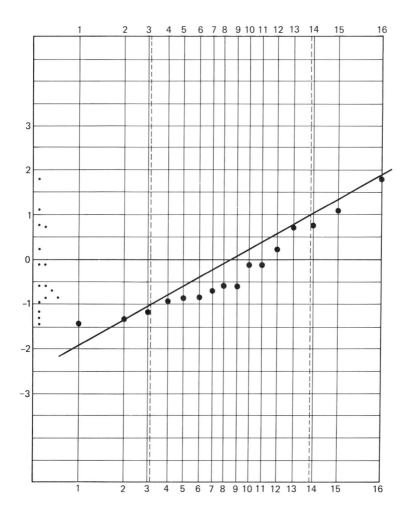

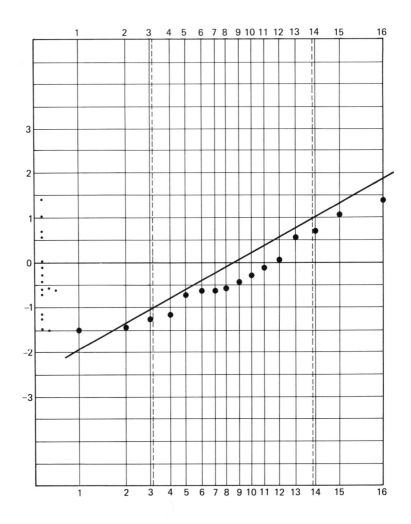

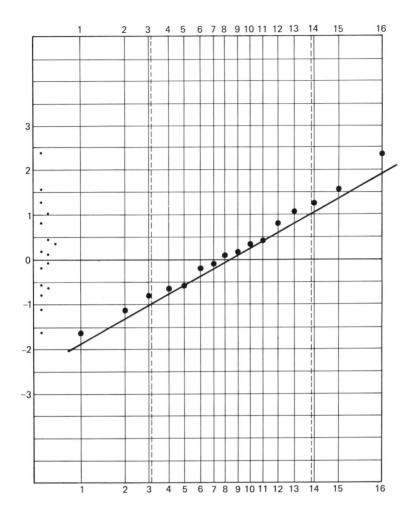

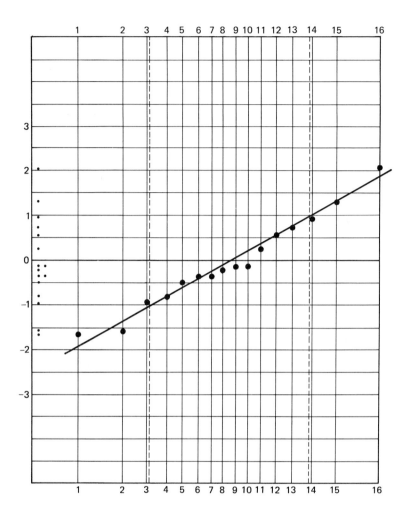

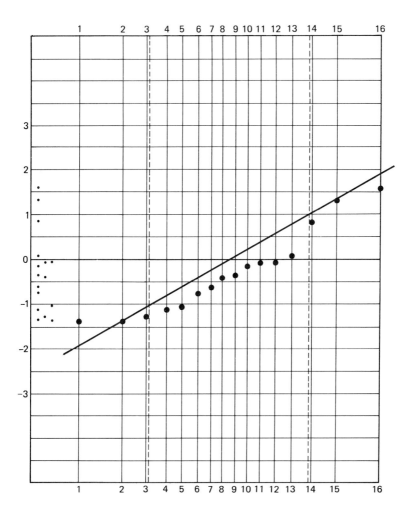

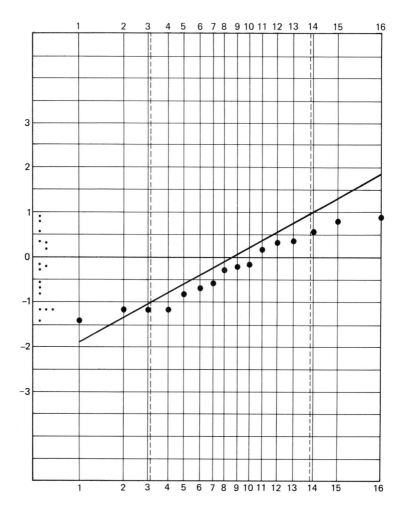

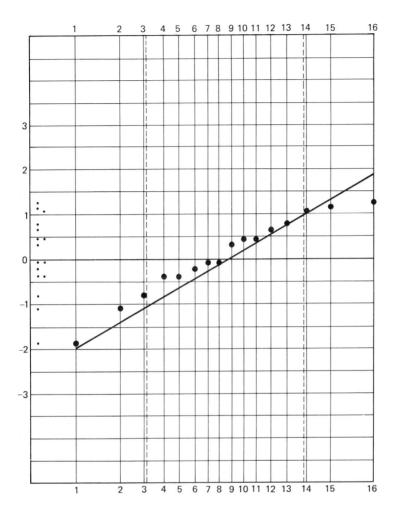

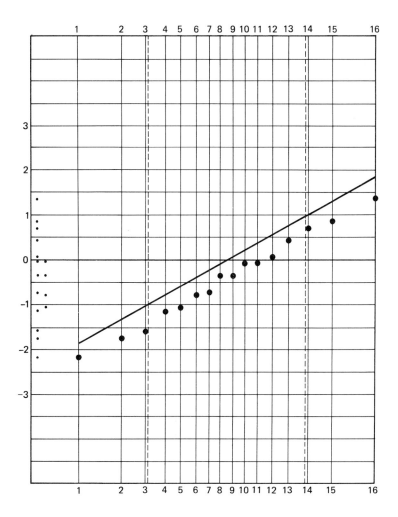

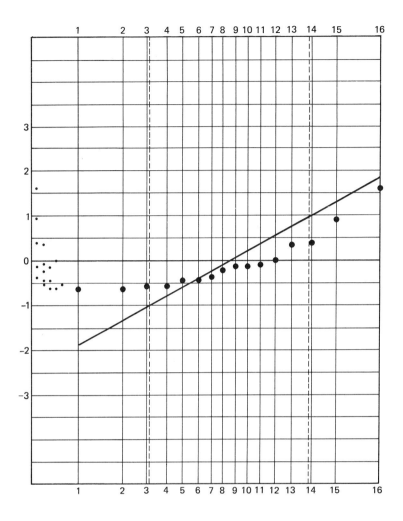

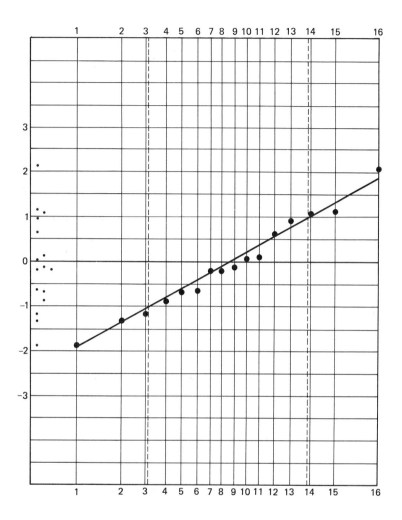

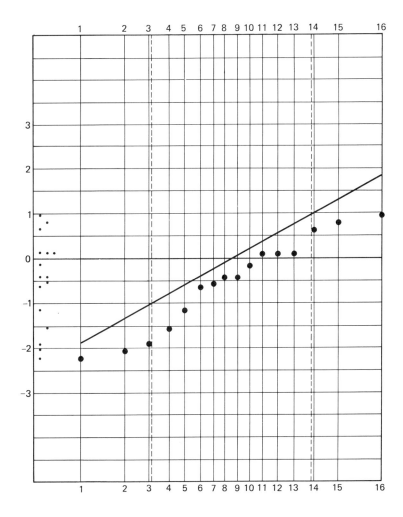

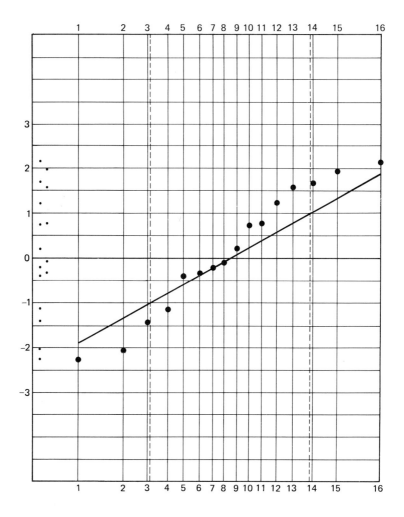

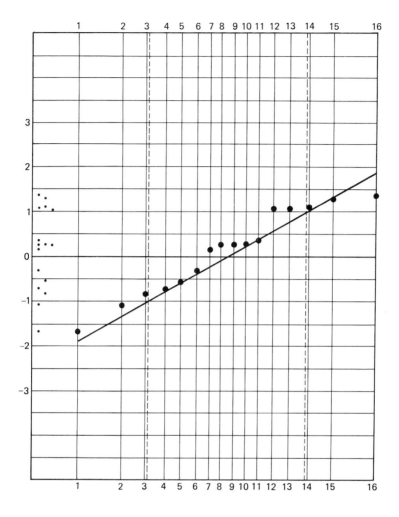

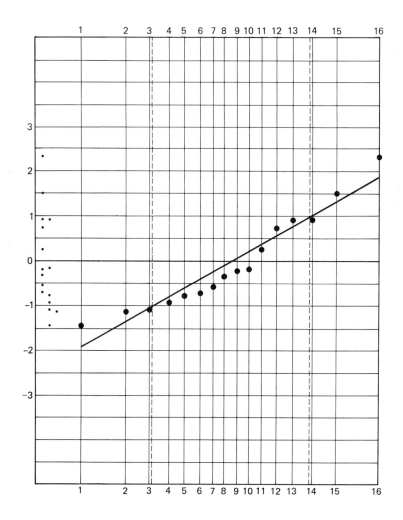

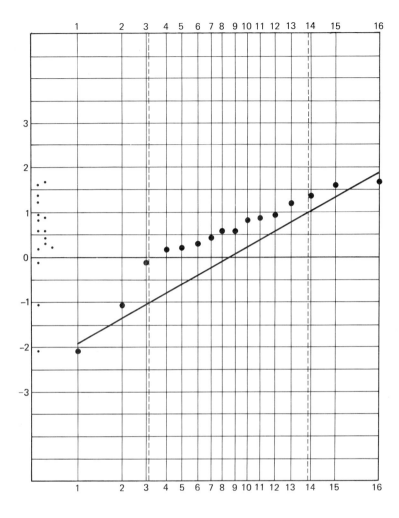

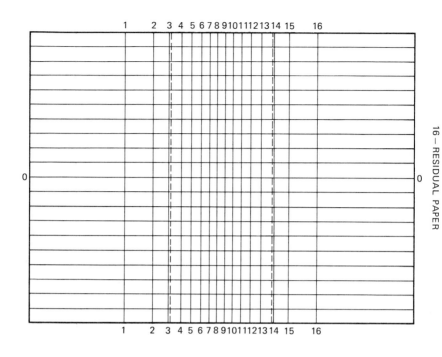

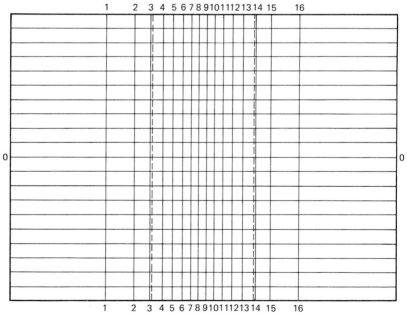

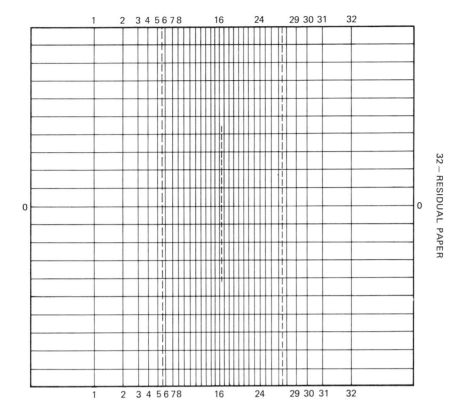

32 – RESIDUAL PAPER

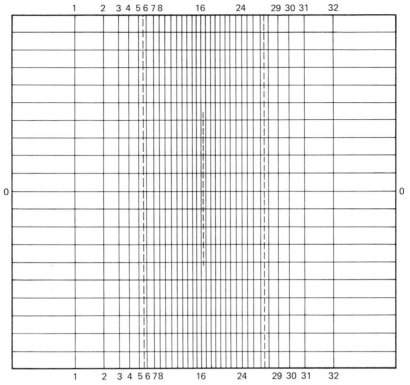

32 – RESIDUAL PAPER

Three Five-Factor, Two-Level Unreplicated Experiments

7.1 INTRODUCTION

Three 2^5 factorials are studied in this chapter. They were not chosen after inspection of a larger set; they are simply the three best-known examples. The program of analysis is like that of earlier chapters but is now augmented.

We expect a small number of real effects and low-order interactions. At most all main effects, some two-factor interactions, possibly a three-factor interaction, and some block effects will be large. We will have, then, about

16 d.f. above and beyond the real effects. These can be used poetically, as degrees of freedom for the imagination. The key assumptions of the standard analysis (uncorrelated errors with constant variance, no block-factor interactions, no bad values) need not remain unexamined. They can sometimes be roughly verified, and sometimes shown to be invalid, by the data. Almost as often, the analysis can be modified to take account of the observed failures of the standard assumptions.

The extra degrees of freedom are sufficiently numerous, as they were not in the smaller experiments discussed earlier, to manifest many identifiable patterns and parts of patterns. The number of contingencies is finite in the sense that a single data analyst will have time to think of only a finite number of things to study, but it is infinite in the sense of being unbounded. This chapter will be successful to the degree that it stimulates readers to study critically the data produced by their own experimental systems.

There will be many failures in that frequently no pattern is detected. Since patterns are often indications of nonrandomness, their absence is desirable, and such failures are welcome.

Some readers will not be able to repress a natural feeling: "If you look long enough, you are bound to find something." Aside from the question-begging term *long enough*, I deny the allegation. I look longer than most, and usually find nothing. But a deeper objection to the statement is that it excuses those who do not look at all, and who believe that the ancient wisdom requiring the analysis to be determined before the data are taken is the only true way. If some of the key assumptions underlying the standard analysis of all factorial experiments can be tested or even refuted by examination of the data, then one who has not noticed this has not looked long enough.

The standard form of the "analysis of variance," which is widely used in summarizing factorial designs with factors at many levels, does not seem to me to be useful for 2^n data. All the contrasts from a 2^n data set must be examined together. Their order, their distribution, and their signs are all lost in the standard analysis of variance table. The habit of summarizing the results in such a table (manifested in so many textbooks that it would be unkind to name) has had a tranquilizing effect with much information lost.

The data juggling that appears in this chapter might be charitably described as a series of efforts to get the data into such shape that they will admit of valid standard significance tests. I do not know the effects of these revisions on the operating characteristics of the tests, and I await the results of further research with fear and trembling.

The three classic experiments are taken from Yates [1937], Davies [1971], and Kempthorne [1952]. They have been studied many times before. Each analysis brings new questions, and no final answers are offered. The work presented here should of course have been carried out in close collaboration with the experimenters, who could have categorically ruled out many of my questions and doubts, and who surely would have raised other more realistic ones.

7.2.   YATES'S 2^5 ON BEANS

7.2.1.   Description

This early experiment, described in Yates's classic pamphlet [1937], was carried out at Rothamsted in 1935. The five factors were as follows: spacing of rows S (18 and 24 in.), amount of dung D (0 or 10 tons/acre), amount of nitrochalk N (0 or 50 lb/acre), amount of superphosphate P (0 or 60 lb P_2O_5/acre), and amount of muriate of potash K (0 or 100 lb K_2O/acre). As Yates writes, "The spacing was varied to test the theory that the best effects of manures might be obtained with closely spaced rows." The field plan is reproduced in Table 7.1. The actual data, arranged in standard order on s, d, n, p, k, are given in Table 7.2.

TABLE 7.1.
FIELD PLAN FOR YATES'S 2^5

	III			IV
nk	(1)		p	npk
snp	sdn		$sdnp$	sn
dp	spk		d	sk
sdk	$dnpk$		dnk	$sdpk$
s	sdp		n	k
snk	dk		$sdnk$	dpk
np	$sdnpk$		sp	dnp
dn	pk		sd	$snpk$
	I			II

The test was arranged in four blocks of eight plots, confounding SDP, SNK, and $DNPK$ with block means. This was accomplished by choosing for the "principal block" [which is III in this case and can always be identified by its containing the treatment combination (1)] those conditions which have an *even* number of letters in common with SDP and with SNK. These can be most expeditiously found by first writing down *three* such combinations which can be used as "generators of the principal block"; the third must not be the product of the other two. Thus dp, nk, and sdn will do. Their products (see Chapter 10) in all combinations, along with (1), give the set used in block III, whose mean is then aliased with $-SDP - SNK + DNPK$. The

TABLE 7.2.
RESULTS OF STANDARD COMPUTATIONS ON YATES'S 2^5 ON BEANS

Spec.	Data (y)	Rounded and -59 (y')	Contrast-Sum	Effect Symbol	Fitting Equation	Y'	d
(1)	66.5	7	-4	T	-4	3	$+4$
s	36.2	-23	-128	S	-128	-26	$+3$
d	74.8	16	252	D	252	6	$+10$
sd	54.7	-4	80			2	-6
n	68.0	9	50			3	$+6$
sn	23.3	-36	54			-26	-10
dn	67.3	8	82			6	$+2$
sdn	70.5	11	30			2	$+9$
p	56.7	-2	-84			-9	$+7$
sp	29.9	-29	48			-14	-15
dp	76.7	18	-8			18	0
sdp	49.8	-9	-188	SDP	-188	-10	$+1$
np	36.3	-23	-82			-9	-14
snp	45.7	-13	18			-14	$+1$
dnp	60.8	2	18			18	-16
$sdnp$	64.6	6	-10			-10	$+16$
k	63.6	5	120	K	120	1	$+4$
sk	39.3	-20	136	SK	136	-10	-10
dk	51.3	-8	-64			5	-13
sdk	73.3	14	-24			18	-4
nk	71.2	12	70			1	$+11$
snk	60.5	1	-98			-10	$+11$
dnk	73.7	15	38			5	$+10$
$sdnk$	92.5	33	-35			18	$+15$
pk	49.6	-9	-8			-10	$+1$
spk	74.3	15	-56			2	$+13$
dpk	63.6	5	-28			17	-12
$sdpk$	56.3	-3	-60			6	-9
npk	48.0	-11	-78			-10	-1
$snpk$	47.9	-11	-102			2	-13
$dnpk$	77.0	18	46			17	$+1$
$sdnpk$	61.3	2	78			-10	-4

treatment combinations for any other block are found by multiplying the eight just obtained by any letter or combination of letters not in the block. As these were allocated to the four blocks shown, we have the confounding:

Block	SDP	SNK	DNPK
I	+	+	+
II	−	+	−
III	−	−	+
IV	+	−	−

7.2.2. Standard Computations

As always, the total and the 31 contrast-sums are computed by Yates's method. The results are given in Table 7.2 for the data rounded to the nearest unit. We see that S, D, SDP (one of the block contrasts), K, and SK are the largest effects. We use these five effects, transferred to the column marked "Fitting Equation" in Table 7.2, in the reverse Yates computation to find the fitted values Y' and thence the residuals d, also given in the table.

The usual plots (empirical cumulative distribution of residuals, and residuals versus Y) are shown in Figures 7.1 and 7.2. These plots show no cause for alarm or even suspicion. The imputed standard deviation of y from Figures 7.1 is 9.8 with 26 d.f., only a little larger than the 9.06 that Yates found (with 13 d.f.) by his more conservative pooling of higher-order interactions not used for blocking.

7.2.3. Residuals in Place

The residuals can also be placed in their respective plot positions, as in Table 7.3. Here we see a region of high fertility in the area inside the dotted line. This region extends into all four blocks, but the largest residuals appear in the two right-hand columns ($+15$, -15, $+16$, -16).

Just as a trial, we look at the effects of the factors in blocks I and III. By good luck, these two blocks comprise a superblock with only $DNPK$ aliased with their mean. Each of the contrasts found from these 16 plots estimates the sum of two aliased effects, but no serious confusion appears. Table 7.4 shows the results of the usual computations, including the pairs of aliased effects and the residuals, from an equation that includes D, K, SK, $DK + NP$, and $SDP + SNK$ (between blocks I and III).

The 16 residuals now have a MS value of 24.7 with 10 d.f., a striking reduction from the 82 found by Yates for the whole 2^5. The reader is spared the necessity of looking at another set of plots and residuals in place. All of them look all right.

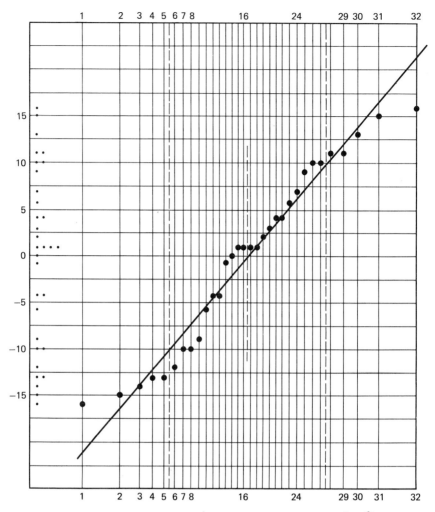

Figure 7.1 Yates's 2^5 on beans. Residuals from $Y(\bar{y}, \hat{S}, \hat{D}, \hat{K}, \widehat{SK}, \widehat{SDP})$.

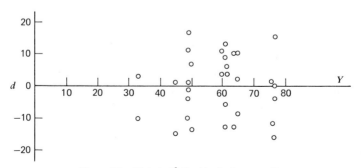

Figure 7.2 Yates's 2^5. Residuals d_j versus Y_j.

TABLE 7.3.
Residuals from 2^5 on Beans, in Place.

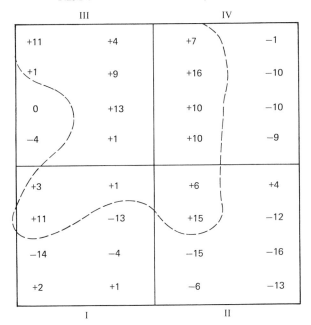

TABLE 7.4.
Standard Computations Using Blocks I and III Only

Spec.	$y - 59$	Contrast-Sum	Effect Symbol	Equation	Y	d
(1)	7	+21	T	+21	0	+7
s	−23	−25	S		−26	+3
dk	−8	+87	$D + NPK$	+87	−7	−1
sdk	14	−11	SD		9	+5
nk	12	+11	$N + DPK$		9	+3
snk	1	−3	SN		−1	+2
dn	8	+37	$DN + PK$		2	+6
sdn	11	−13	SDN		12	−1
pk	−9	−23	$P + DNK$		−8	−1
spk	15	+7	SP		17	+2
dp	18	+31	$DP + NK$		20	−2
sdp	−9	−143	$SDP + SNK$	−143	−6	−3
np	−23	−73	$NP + DK$	−73	−18	−5
snp	−13	−3	$SNP + SDK$		−8	−5
$dnpk$	18	+69	$DNP + K$	+69	11	+7
$sdnpk$	2	+63	$SDNP + SK$	+63	1	+1

7.2.4. Dropping the Factorial Representation

The complex of large main effects and 2fi's, all involving S, D, and K, may be trying to tell us something, but they are not saying anything simple in the factorial representation. We may have another case before us—they are not rare—in which Nature is not behaving "factorially," at least not in terms of our present factors. Since the four effects, D, K, SK, and DK, are of roughly the same magnitude (87, 69, 63, and -73, respectively), we set them all at $+1$ or -1 and put them through a reverse Yates computation in Table 7.5. The resulting eight (coded) fitted values are placed on a 2^3 diagram in Figure 7.3. A 2^3 suffices since N and P are without effects.

Although we do not see any simple way to summarize these results, something should be said. From the whole 2^5 Yates had found, concerning the

TABLE 7.5.
JUDGING THE SIMULTANEOUS IMPACT OF
D, K, SK, DK

	(0)	(1)	(2)	(3)	
SDK	0	-1	1	2	$\widehat{sdk}$
DK	-1	2	1	0	$\widehat{dk}$
SK	1	1	-1	2	$\widehat{sk}$
K	1	0	1	0	$\hat{k}$
SD	0	-1	3	0	$\widehat{sd}$
D	1	0	-1	2	$\hat{d}$
S	0	1	1	-4	$\hat{s}$
T	0	0	-1	-2	$(\hat{1})$

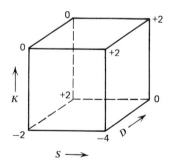

Figure 7.3   Coded responses to S, D, and K in Yates's 2^5 on beans.

earlier conjecture that the best effects might be obtained at closer spacing, that "the interaction between spacing and manures turned out to be the opposite of what had been expected." If the results of these two blocks are to be believed, then dung alone on closely spaced rows (i.e., treatment combination d) is as good as any other combination found. Indeed at wide spacing potash helps, as sk shows, but adding dung after (or before?) potash, as at sdk, gives no further increase.

There are, then, three favorable combinations, and they are of rather unexpected variety—d, sk, and sdk. We suppose that no one would add potash and dung if either alone did as well, and we presume that wide spacing would be easier to manage and cultivate. This puts sk ahead. Not surprisingly, the worst combination is wide spacing and *no* fertilizer of any kind (i.e., s). Finally, as is usually the case in this country too, fertilizer nitrogen does not appear to increase the yield of beans.

7.2.5. A COMMON RESULT: $|A| \doteq |B| \doteq |AB|$

Returning for a moment to the full 2^5 shown in Table 7.2, we see that the total effects for S, K, and SK are -128, 120, and 136, respectively. This rough equality turns up quite frequently and has a simple interpretation. If we put the three effects at -1, 1, and 1 for simplicity, and go through a 2^2 reverse Yates computation, we have the following:

		S	
		0	3
K	0	1	−3
	1	1	1

There are, then, three nearly identical combinations, and one exceptional one, here s. (Other assortments of signs of the three effects simply move the exceptional response to some other cell or change its sign.) The simple interpretation of such an equality of magnitude of A, B, and AB is: One combination is exceptionally high (or low), and the other three combinations produce the same average response.

7.3. DAVIES' 2^5 ON PENICILLIN [1971, PAGES 383, 415]

7.3.1. Description

This was a full 2^5 done in two blocks (weeks) to study the effects on yield of variation in the concentrations of five components of a nutrient medium for growing *Penicillium chrysogenum* in surface culture.

Table 7.6 gives the yields y and their coded logs, $z = 100(\log y - 2)$, together with the usual contrasts. We see by inspection that the influential factors must be A, C, and E, with perhaps CE and blocks ($ABCDE$) having some influence. This is true both for y and for log y. We note that the per cent of the total sum of squares accounted for by A, C, E, and CE is 86.5 for y, and 76.2 for z.

TABLE 7.6.

DAVIES' 2^5 ON PENICILLIN: $y' = $ yield $- 130$; $z = 100(\log$ yield $- 2)$

Spec.	y'	Effects on y'	Name	z	Effects on z	Revised Effects on y'	Y'	$d_y{}^\dagger$	Effects on $\lvert d_y \rvert$
(1)	12	-14^*	T	15	304	61^*	9	$+3$	
a	-16	-562^*	A	6	-190	-487^*	-21	$+5$	-27
b	-1	18	B	11	-6	93	9	-10	61
ab	-21	-190	AB	4	-64	-115	-21	0	-59
c	55	514^*	C	27	153	583^*	72	-17	1
ac	32	-194	AC	21	-53	-119	41	-9	13
bc	70	142	BC	30	53	217	72	-2	25
abc	42	-42	ABC	24	0	33	41	$+1$	17
d	18	32	D	17	9	107	9	$+9$	-1
ad	-22	-160	AD	3	-54	-85	-21	-1	79
bd	16	-32	BD	16	-7	43	9	$\vert 7$	23
abd	-35	-92	ABD	-2	-34	-17	-21	-14	27
cd	70	-32	CD	30	-4	43	72	-2	-49
acd	34	56	ACD	22	33	131	41	-7	31
bcd	85	-84	BCD	33	-18	-9	72	$+13$	35
$abcd$	-12	128	$ABCD$	7	58	203	41	$+22$	-49
e	-24	-668^*	E	2	-224	-743^*	-12	-12	157^*
ae	-24	84	AE	2	2	9	-42	$+18$	-19
be	-42	96	BE	-6	29	21	-12	-30	29
abe	-32	-52	ABE	-1	-22	-127	-42	$+10$	-87
ce	-17	-336^*	CE	5	-93	-411^*	-1	-16	-47
ace	-42	-104	ACE	-6	-58	-179	-31	-11	-15
bce	36	88	BCE	22	39	13	-1	$+37$	45
$abce$	-51	84	$ABCE$	-10	31	9	-31	-20	1
de	-29	70	DE	0	30	-5	-12	-17	-57
ade	-16	90	ADE	6	21	15	-42	$+26$	19
bde	10	74	BDE	15	28	-1	-12	$+22$	-93
$abde$	-58	58	$ABDE$	-14	14	-17	-42	-16	-21
cde	0	18	CDE	11	12	-59	-1	1	-53
$acde$	-47	134	$ACDE$	-8	47	59	-31	-16	-1
$bcde$	-15	34	$BCDE$	16	16	-41	-1	$+16$	-45
$abcde$	-20	202^*	$ABCDE$	4	77	127	-31	$+11$	11

* Judged significant.

† Results after revising $abcd$ by $+75$.

7.3.2.   When to Log

The authors write, "The logarithmic transformation was used because the error was expected to be proportional to the result." The basic reason for "logging" the dependent variable must be that the equation representing the data is expected to be of the form

$$(7.1) \qquad\qquad Y = a \exp (bx_1 + cx_2 + dx_3 + \cdots)$$
$$= aC^{x_1}D^{x_2}E^{x_3} \cdots,$$

where $C = \ln b_1$, etc.

Thus the factors are expected to operate exponentially on the response. It may well be that the uncontrolled factors, which are producing the random variation in y, also operate exponentially. But even if there were no error at all, the logarithmic transformation of y would be obligatory if the system followed an equation like (7.1) for its influential factors. In the present case overall per cent error appears to be about 13% from Davies' analysis. For coefficients of variation (i.e., per cents error) less than 20% the random error will usually not be useful in deciding on logging. In these data, most of the variation is being produced by the systematically varied factors, not by the random error. It is the former, then, not the latter, that will give more information about transforming.

At this point we would choose y, not z.

7.3.3.   A Bad Value

There are 13 effects (contrasts) in the *latter* 16 that should be error estimates. I exclude E, CE, and $ABCDE$ since the former two are plausible effects and the last is a blocking contrast. Of these 13, there are 11 that are positive. A discrepancy from evenness as large as this would occur only with relative frequency 0.01123 (National Bureau of Standards, *Tables of Binomial Probability Distribution*, 1950, page 211, $p = .50$, $n = 13$, $r = 11$). Since an equal discrepancy in the opposite direction would be equally striking, I find a tail probability of 0.02246.

Of the 13 candidates for random error in the *upper* 16 effects of Table 7.6 (here I exclude T, A, and C) I count 9 in the expected negative direction, and hence 5 (namely, AB, BC, D, ACD, and $ABCD$) are in the adverse direction. There are, then, 7 out of 26 adverse, 19 in favor. A divergence as large as this in either direction from 13:13 is of probability .02896 (*ibid.*, page 252, $p = .50$, $n = 26$, $r = 19$), and so is still quite unlikely.

There is only one response that can bias the contrasts in this way, and it is *abcd*. Looking at the magnitudes of the lower set, I guess that *abcd* must be off by about 75. If I revise the value of y at *abcd* to $+63$, I get a new set of total effects, each differing by ± 75 from the former one. The entire revised set is shown in Table 7.6, along with the new residuals found by using the revised effects A, C, E, and CE. The block difference, $ABCDE$, has now dropped to an inconsequential level.

7.3.4.   Effects on Residuals

We now have a set of data that looks "all right," but an upsetting aspect appears in the residuals as ordered: those at high E are clearly larger than those at low. We cannot count on being so fortunate in the future as we are here, where the unstabilizing factor appears to be E and so is clearly visible on inspection. We might take the absolute values of the residuals and put *them* through the Yates procedure. The result is seen in the last column of Table 7.6. Not surprisingly, factor E emerges clearly separated from all other "effects on the magnitude of $|d_y|$."

We are forced to study the 2^5 as two 2^4's. Table 7.7 shows the usual results, obtained first *without* revision of $abcd$. In panel a for low E we see that in this context a revision by about $+50$ will suffice. The table shows the results of this revision, both on the effects and on the residuals from the obvious fitting equation in A and C. The MS residual is now 77.2; the estimated standard deviation (std. dev.), 8.8. This corresponds to a 6% precision and an R^2 of 0.953, both welcome changes from the 13% precision and R^2 of 0.865 that we might have reported for the whole 2^5.

The 16 results at high E present a much less satisfactory picture. Panel b of Table 7.7 gives the results of the routine computations. No bad values are obvious in this set. The std. dev. is 19, which, with the mean of 109, gives a precision of 17%. The two effects detected are A and $ABCD$, which is aliased with the block difference in this half replicate. These two effects give an R^2 of 0.524.

Davies writes, "Information existed from earlier work that interaction CE was likely to be appreciable." This can now be given fuller justification and a simpler statement: Factor C was strongly influential at low E, and not at all at high.

We must reject the set of high-E runs as having too great a variance, as showing a block effect not present at low E, and as not being of practical importance anyway since E has a large adverse effect on yield. We may take some comfort from the fact that the improved precision of the low-E set permits us to estimate the effects of A and of C as precisely as they were apparently estimated from the whole 2^5.

7.3.5.   Conclusions

Only the low-E half of this 2^5 is safely interpretable. It has excellent precision (6%), one bad value (at $abcd$), and two clear main effects, A and C. The other half has poor precision (17%) and one clear effect, A. But since

TABLE 7.7.
DAVIES' 2^5 AS TWO 2^4's, ONE AT LOW E AND ONE AT HIGH E

Panel a. y = yield $- 150$

Specs.	y	Effects on y	Name	Revised Effects[†]	$Y^†$	$d_y{}^†$	
(1)	-8	7	T	57	-9	$+1$	
a	-36	-323	A	-273^*	-43	$+7$	
b	-21	-39	B	11	-9	-12	
ab	-41	-69	AB	-19	-43	$+2$	
c	35	425	C	475^*	50	-15	
ac	12	-45	AC	5	16	-4	
bc	50	27	BC	77	50	0	
abc	22	-63	ABC	-13	16	$+6$	
d	-2	-19	D	31	-9	$+7$	Std. Dev. (d) = 8.8.
ad	-42	-125	AD	75	-43	$+1$	
bd	-4	-53	BD	-3	-9	$+5$	R^2 = 0.953.
abd	-55	-75	ABD	-25	-43	-12	
cd	50	-25	CD	25	50	0	
acd	14	-39	ACD	11	16	-2	
bcd	65	-59	BCD	-9	50	$+15$	
$abcd$	-32	-37	$ABCD$	13	16	$+2$	

Panel b. y = yield $- 109$

e	-3	-5			25	-28	
ae	-3	-239^*	A		-25	$+22$	
be	-21	57	B		2	-23	
abe	-11	-121			-2	-9	
ce	4	89	C		2	$+2$	
ace	-21	-149			-2	-19	
bce	57	115			25	$+32$	
$abce$	-30	21			-25	-5	
de	-8	51	D		2	-10	
ade	5	-35			-2	$+7$	Std. Dev. (d) = 19.2.
bde	31	21			25	$+6$	
$abde$	-37	-17			-25	-12	R^2 = 0.524.
cde	21	-7			25	-4	
$acde$	-26	95			-25	-1	
$bcde$	36	-25			2	$+34$	
$abcde$	1	165^*	$ABCD$ + blocks		-2	$+3$	

* Judged real.
† $abcd$ revised by $+50$.

high E itself is unfavorable to yield, we have lost the less useful half of this experiment.

More generally, statisticians have too long applied a weak criterion for deciding whether to use y or log y. The decisive criterion is: Which form gives the better representation of the data, not of its error distribution? We must not let the tail wag the log.

We can get clues to the dependence of the magnitude of the error on experimental conditions by putting the absolute values of the residuals through the usual process for two-level factorials.

A simple bad value can be spotted before commitment to a set of effects judged real. There are certain patterns (2×2^n) in the ordered contrasts that indicate just which value is biased and in which direction the bias lies.

7.4. ROTHAMSTED'S 2^5 ON MANGOLDS
(COPIED FROM KEMPTHORNE)

7.4.1. Description

The data are taken from Section 14.7 of Kempthorne [1952], pages 267–270. The five factors—S, P, K, N, D—were amounts of sulfate of ammonia, superphosphate, muriate of potash, agricultural salt, and dung, respectively. Each was varied from *none* to some. The experiment was divided into four blocks, all in a single field. Table 7.8 gives the actual field arrangement and yields.

TABLE 7.8.
FIELD PLAN AND YIELDS FOR ROTHAMSTED'S 2^5: FROM KEMPTHORNE [1952]

I				II			
pkd	*nd*	*sk*	*spknd*	*d*	*pknd*	*k*	*snd*
844	1104	1156	1506	1248	1100	784	1376
spn	*kn*	*sd*	*p*	*spkd*	*skn*	*sp*	*pn*
1312	1000	1176	888	1356	1376	1008	964
kd	*spd*	*pnd*	*pkn*	*skd*	*spkn*	*knd*	*spnd*
896	1284	996	860	1328	1292	1008	1324
sn	*spk*	*(1)*	*sknd*	*pd*	*pk*	*n*	*s*
1184	984	740	1468	1008	692	780	1108
III				IV			

The confounding pattern can be found expeditiously by setting down the *three* generators (any three) of the principal block (here III), and then finding the interactions that have an *even* number of letters in common with each of these generators. Thus, if we choose *sn*, *kd*, and *spd* as generators, we find by direct trial that $-SPN$, $-PKD$, and so necessarily their product, $+SKND$, are confounded with the mean of block III (see Chapter 10). We note that SPN is $+$ in blocks I and IV and $-$ in blocks II and III, and that PKD is $+$ in I and II and $-$ in III and IV. So, necessarily, $SKND$ is $+$ in I and III. We will want to remember that PKD is confounded with the "north" versus "south" pairs of blocks, that $SKND$ is confounded with the "east-west" distinction, and that SPN is confounded with the difference between the diagonal pairs, that is, $(I + IV) - (II + III)$.

7.4.2. Standard Computations

The usual 31 contrast-sums are computed in Kempthorne's Table 14.11. (The yield at *pn* should be 964, not 864.) Since we know that the error std. dev. is about 80, there can be no harm in rounding the observations to the nearest 10. Table 7.9 shows as y (column 3) the yields in units of 10 lb, in standard order. The next column, headed y', gives $(y - 110)$. The next column lists the resulting contrast-sums. These are shown to demonstrate how closely they match those given by Kempthorne and to simplify future reference.

Since S, N, and D are visibly controlling, I use only these three at first to construct a fitting equation and to find the residuals, d_1. The fitted values are produced from the fitting equation in Table 7.10, and the residuals are placed in standard order in Table 7.9. They are plotted as an *ecd* in Figure 7.4. We hardly need a test for outliers to justify studying the effect of the observation at d on the estimate of error and on the choice of "real" effects.

7.4.3. One Bad Value?

The MS residual computed from the d_1 is $2720/28 = 97.2$. We can find the change in each residual—when the value at d is revised by an amount Q—without recomputing all effects. A change of any one observation by the amount Q will change every contrast-sum by $\pm Q$. Here we are interested in only three effects, namely, S, N, and D and in the mean. The signs can be read from Davies' Table M or can be written down directly. Since we are considering a *decrease* in the value at d, we know that this will appear *positively* in S and in N, and negatively in D and in T. The four values, most simply set at $+1$, $+1$, -1, and -1, respectively, are run through reverse Yates (see Table 7.10, panel b) and then rescaled in units of Y by multiplication by $d_1/2^p$. In the present case $d_1/2^p$ is $29/32$, which is so close to 1 that no rescaling is needed. I apologize for the inapt use of the symbol Δ for a change in a statistic and not, as is usual, for a parameter.

TABLE 7.9.
ROTHAMSTED'S 2^5: EFFECTS, FITTED VALUES, RESIDUALS

(1) Block	(2) Specs.	(3) y	(4) y'	(5) Effects	(6) Real	(7) Y_1'	(8) d_1	(9) Δd_1	(10) d_2
III	(1)	74	-36			-32	-4	$+2$	-2
IV	s	111	1	534	534S	1	0	0	0
I	p	89	-21	-34		-32	$+11$	$+2$	$+13$
II	sp	101	-9	8		1	-10	0	-10
II	k	78	-32	16		-32	0	0	0
I	sk	116	6	126		1	$+5$	$+2$	$+7$
IV	pk	69	-41	-46		-32	-9	0	-9
III	spk	98	-12	-8		1	-13	$+2$	-11
IV	n	78	-32	214	214N	-19	-13	0	-13
III	sn	118	8	72		14	-6	-2	-8
II	pn	96	-14	44		-19	$+5$	0	$+5$
I	spn	131	21	-10		14	$+7$	-2	$+5$
I	kn	100	-10	102		-19	$+9$	0	$+9$
II	skn	138	28	-60		14	$+14$	-2	$+12$
III	pkn	86	-24	-4		-19	-5	0	-5
IV	$spkn$	129	19	10		14	$+5$	-2	$+3$
II	d	125	15	292	292D	-14	$+29$	-29	0
I	sd	118	8	-10		19	-11	$+2$	-9
IV	pd	101	-9	-6		-14	$+5$	$+4$	$+9$
III	spd	128	18	76		19	-1	$+2$	$+1$
III	kd	90	-20	-16		-14	-6	$+4$	-2
IV	skd	133	23	78		19	$+4$	$+2$	$+6$
I	pkd	84	-26	126	PKD	-14	-12	$+4$	-8
II	$spkd$	136	26	-60		19	$+7$	$+2$	$+9$
I	nd	110	0	-66		-1	$+1$	$+2$	$+3$
II	snd	138	28	-8		32	-4	0	-4
III	pnd	100	-10	-16		-1	-9	$+2$	-7
IV	$spnd$	132	22	-78		32	-10	0	-10
IV	knd	101	-9	14		-1	-8	$+2$	-6
III	$sknd$	147	37	-36		32	$+5$	0	$+5$
II	$pknd$	110	0	40		-1	$+1$	$+2$	$+3$
I	$spknd$	151	41	22		32	$+9$	0	$+9$

TABLE 7.10.
COMPUTATION OF Y_1 AND EFFECTS ON d_1

Panel a. Y_1

	(0)	(1)	(2)	(3)	(3) ÷ 32	
SND		0	292	1040	32	$\widehat{snd}$
ND	292	748	−28	−1		$\widehat{nd}$
SD	214	292	612	19		$\widehat{sd}$
D	292	534	−320	−456	−14	$\hat{d}$
SN		0	202	456	14	$\widehat{sn}$
N	214	292	320	−612	−19	$\hat{n}$
S	534	214	292	28	1	$\hat{s}$
T	0	−534	−748	−1040	−32	$(\hat{I})$

OR

	(0)	(1)	(2)	
SN − D	−292	−78	456	$\widehat{sn} = -\hat{d}$
N − SD	214	534	−28	$\widehat{nd} = -\hat{s}$
S − ND	534	506	612	$\widehat{sd} = -\hat{n}$
T − SND	0	−534	−1040	$(\hat{I}) = -\widehat{snd}$

Panel b. d_1

	(0)	(1)	(2)	(3)	
SND		0	−1	2	$\widehat{snd}$
ND		−1	3	0	$\widehat{nd}$
SD		1	−1	0	$\widehat{sd}$
D	−1	2	1	−2	$\hat{d}$
SN		0	−1	4	$\widehat{sn}$
N	1	−1	1	2	$\hat{n}$
S	1	1	−1	2	$\hat{s}$
T	1	0	−1	0	$(\hat{I})$

The changes in d_1, designated as Δd_1, will be the negatives of those in Y, and these values are transferred to Table 7.9. The revised residuals d_2 are also shown. Their MS is 1755/27 or 65.00, which is close to the 67.91 given by Kempthorne with 13 d.f.

7.4.4. Dependence of Residuals on Y or X?

Figure 7.5 shows no visible tendency of the d_2 residuals (obtained after revision of the response at d by -29) to vary with Y. We have put the absolute values of the d_2 through Yates's algorithm to see whether there is any dependence of the sizes of the residuals on experimental conditions. Since nothing striking emerges, we have not reproduced the results.

7.4.5. Block-Factor Interactions?

The contrast labeled PKD in Table 7.9 appears large and has a simple alias. It is the difference between the total yield in the upper pair of blocks and that in the lower pair. We ask, then, whether the effects of factors S, N, D are the same in these two pairs.

Two more standard computations, one on the "half replicate" defined by $I + PKD$, the other on the half $I - PKD$, give the total effects listed in Table 7.11 (see Chapter 11). The effects do not seem the same. It appears

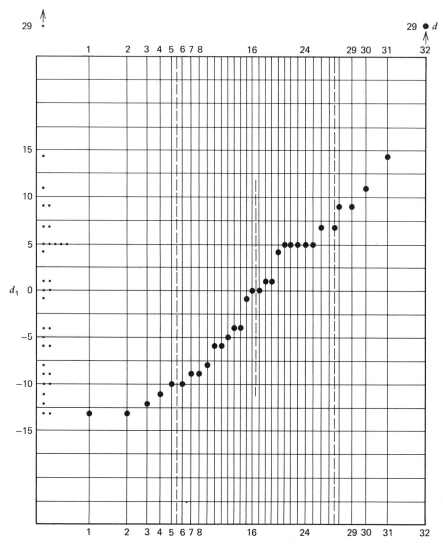

Figure 7.4 Rothamsted's 2^5. Residuals d_1 from $Y_1(\bar{y}, \hat{S}, \hat{D}, \hat{N})$.

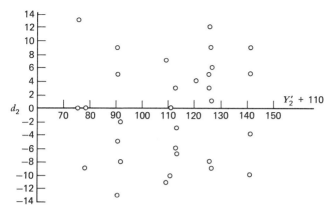

Figure 7.5 Rothamsted's 2^5. Residuals d_2 versus Y_2 (one value revised).

TABLE 7.11.

FACTORIAL EFFECTS NORTH (I AND II) AND SOUTH (II AND IV)

Blocks I and II: $I + PKD$		y'	Contrasts		d_3	Blocks III and IV: $I - PKD$		y'	Contrasts	d_3
(1)	k	-32	---		-2	(1)		-36	---	$+3$
s	k	6	267*	S	$+3$	s		1	297*	$+3$
p		-21	5	P	$+9$	p	k	-41	-9	-2
sp		-9	13		-12	sp	k	-12	-35	-10
n	k	-10	157*	N	$+1$	n		-32	87*	-4
sn	k	28	17*		-6	sn		8	25	-1
pn		-14	-1		-3	pn	k	-24	15	$+4$
spn		21	7		-2	spn	k	19	13	$+10$
d†		-15	93*	D	$+4$	d	k	-20	169*	-2
sd		8	21		-7	sd	k	23	-1	$+3$
pd	k	-26	35	K	-7	pd		-9	-11	$+9$
spd	k	26	71		$+11$	spd		18	-25	-2
nd		0	-5		-1	nd	k	-9	-31	-2
snd		28	-29		$+5$	snd	k	37	-9	$+7$
pnd	k	0	13		-1	pnd		-10	-59	-3
$spnd$	k	41	-39		$+7$	$spnd$		22	-9	-8

* Judged real.

† Revised from $+15$.

that D has a larger effect in the lower, less fertile pair of blocks, and that N, on the contrary, has a larger effect in the upper pair.

Using the *six* observed effects to derive new residuals, we obtain the values given as d_3 in Table 7.11. These are plotted as a combined e.c.d. in Figure 7.6 and are shown in place in Table 7.12. Lacking any two-dimensional run or cluster theory, I remark only that there appears to be a ridge of high fertility running north and south in the middle of the field.

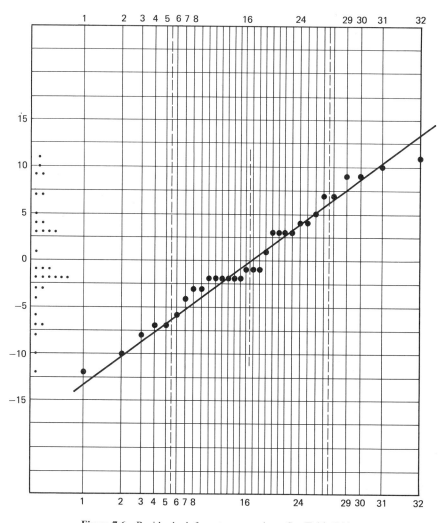

Figure 7.6   Residuals d_3 from two equations. See Table 7.11.

TABLE 7.12

RESIDUALS d_3 IN PLACE ON FIELD PLAN

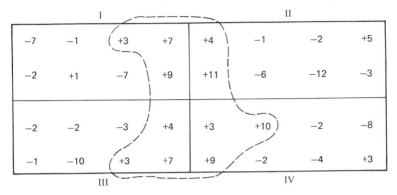

Computing an error MS from the d_3 with 23 d.f. (32 observations $-$ 2 means $-$ 6 effects $-$ 1 bad value), we find 1078/23 or 46.8, and so an estimated error std. dev. of 6.8, checking closely enough with the 6.6 estimated from Figure 7.6. The MS is 70% of that found by Kempthorne with 13 d.f.

7.4.6. Conclusions

We have found rather strong evidence that the factors S, N, and D operated additively in the north pair of blocks and in the south pair, but with different effects. The random error is (of course) considerably reduced from that found earlier, and was due largely to a ridge of high fertility in the middle of the field.

7.5. GENERAL CONCLUSIONS

These studies of ancient 2^5 experiments should not be taken as destructive exercises. Although our results cannot now be useful to the original experimenters, they should be suggestive to the reader now working on large 2^p factorials.

Many experimenters have instinctively—and rightly—resisted turning over their data to a statistician for "analysis," that is, for routine analysis. They have resisted, I believe, because they did not know what the statistician was doing, what unstated assumptions he was making, or what many of his technical terms meant. Many statisticians have been obscure in describing their operations; many have not inquired of the experimenter about his private assumptions; most have referred experimenters to textbooks for definitions of terms.

In this chapter I have tried to show how some of the key assumptions about the system under study can be checked when the data from a 2^5 are at hand.

In most cases the analysis of the data can be suitably modified when some of the assumptions are found inapplicable.

Let us review once more the assumptions of the *experimenter* and those of the *statistician*. I believe that the former nearly always assumes the following:

1. Some of the factors varied will have large and uniform, that is, additive, effects.
2. Some factors will have negligible effects or none at all.
3. A few factors may operate nonadditively. Thus one factor may be quite influential at *one* level of another factor, but much less influential, or even without effect, at another level of the second factor.
4. Sets of data taken close together—blocked—will have small random error for internal comparisons, but *the same effect for each factor should show in each block.*
5. There may be a few wild values, caused by mistakes or by factors other than those deliberately varied.

To these assumptions the statistician will want to add two others:

6. The random errors operating may be viewed as uncorrelated from a single distribution, roughly normal, with zero mean and all with the same variance. To "guarantee" the validity of this assumption, the treatment combinations must be assigned to the plots or experimental units at random.
7. The experimental plots are a fair sample of those to which it is desired to generalize.

To particularize these generalities, see Table 7.13. We note the frequent failure of several key assumptions in these by no means exceptional examples.

TABLE 7.13.
SUMMARY OF RESULTS OF CHAPTER 7

	Assumption	Yates	Davies	Kempthorne
1.	Few large effects	S, D, K	A, C, E	S, N, D
2.	Some negligible effects	N, P	B, D	P, K
3.	Few interactions	SK	CE	None
4.	*a.* Block differences	Yes	Yes	Yes
	b. No block interactions	Large	Present	Large
5.	No wild values	None	*abcd* low	*d* high
6.	Normal error	Yes, in 2 blocks	Yes, in 2 blocks	Yes
7.	Homogeneous, typical plots	High and low strips; see Table 7.3	---	North-south ridge; see Table 7.12

All that remains is to point out a new virtue of factorial 2^p experiments. Just as these plans are sensitive to certain defects—especially to bad values, to large differences in variance of groups of observations, and to block-effect interactions—so their high degree of articulation makes it easier to find these defects. As Yogi Berra is said to have said, "You can observe a lot by just watching."

7.6. HALF-NORMAL PLOTS

I have of course made half-normal plots [Daniel 1959] of the contrast-sums from each of the 2^5 experiments discussed in this chapter. To my great chagrin none of the peculiarities discovered above is reflected in these plots, nor have any other notable irregularities been found. The reasons are not far to seek. The defects found are all strongly sign dependent, and all are properties of subsets of the data set which are obscured in the half-normal plots by overaggregation.

The signed contrasts *in standard order* have more information in them than do the unsigned contrasts ordered by magnitude. The signed residuals from a fitting equation made from the largest effects will often tell us if something is awry. Repeated trials, subdividing the data in plausible ways and inspecting residuals from each trial, will sometimes reveal just what the trouble is.

These failures of half-normal plots prompt Sermon IV: Do not ever assume that a statistic aggregated over a whole data set is distributed as required by some unverified assumptions. The homogeneity of the *parts* of an aggregate can be tested before or after the aggregation, but such testing must be done before conclusions can be drawn from the experiment.

CHAPTER 8

Larger Two-Way Layouts

8.1. INTRODUCTION

We return now to two-way layouts of larger size than the 3 × 3 discussed in Chapter 4. Some new devices for analysis emerge, justified by the higher cost of such data and by the more detailed information they supply. The data sets used as examples range from 3 × 5 to 33 × 31.

The standard factorial representation, including nonadditivity parameters, is less objectionable in the larger data sets. But it is found repeatedly that interactions are localized, not spread irregularly over whole tables. Large interactions appear most often in single cells, in a single row (or column), or occasionally in a few rows or columns. No clear examples of widely disseminated interaction are evident.

We show in detail why the usual residuals behave more nearly like unbiased estimates of true interactions in the larger tables, and we make suggestions for separating these interactions from random error, even in unreplicated data. When the data are as large a set as an 8 × 5, it becomes possible to learn something about the form of the error distribution. Even row-wise heterogeneity of error begins to be detectable.

Large, properly replicated $R \times C$ tables are rare in industrial and physical research, probably because experimenters have recognized that they do not usually need replication for secure interpretation. Although it must be admitted that unrandomized data cannot be as securely interpreted, there is still a lot of room between full security and worthlessness.

151

Some large two-way layouts have quantitative, even equally spaced, levels of both factors; most of the others have qualitative, usually not orderable, factors both ways. J. Mandel is the unrivaled master of model making and data fitting for the first type, and any exposition here would be only an echo of his work [1971; 1964, Chapter 11; 1969a, b]. We concentrate therefore on the unordered discrete-leveled $R \times C$ case.

The effort required to complete a large $R \times C$ table, and the experience that main effects dominate—with occasional interaction cells or rows—combine to suggest that some form of fractionation might be appropriate. Section 8.8 is devoted to this situation.

8.2. A 7 × 4 FROM YATES [1970, PAGE 55]

The data come from a comparative trial of four varieties of cotton (in 4×4 Latin squares) at each of seven centers. "Log yields were tabulated because of evidence of rough proportionality of means to standard errors."

TABLE 8.1.

YIELDS OF 4 VARIETIES OF COTTON AT 7 CENTERS YATES [1970, PAGE 55]
The entries y in the table are related to Yates's values y' as follows: $y = 100(y' - 0.49)$.

Panel a					Sums	Averages			Residuals	
		Data								
-30	-39	-34	-24		-127	-31.8	1.8	-4.5	2.8	0.0
4	5	-4	10		15	3.8	.2	3.9	-2.8	-1.6
-20	-21	-25	-13		-79	-19.8	$-.2$	1.5	-0.2	-1.0
53	47	49	76		$+225$	56.2	-3.2	-6.5	-2.2	(12.0)
-33	-31	-34	-26		-124	-31.0	-2.0	2.7	2.0	-2.8
8	6	0	14		28	7.0	1.0	1.7	-2.0	-0.8
16	12	11	16		55	13.8	2.2	0.9	2.2	-5.6

Sums:	-2	-21	-37	53	-7	Residual MS $= 335/18 = 18.6.$
Sums + 2:	0	-19	-35	55	1	SS (4, 4) $= 12^2 \times 18/12 = 224.$
Deviations:	0	-2.7	-5.0	7.8		Remaining MS $= 111/17$
						$= 6.53.$
						MNR $= 12/335^{1/2} = 0.656.$
						$P \ll .01.$

Panel b. Propagation of a disturbance of RC in cell $(1, 1)$ throughout an $R \times C$ table: $r = R - 1, c = C - 1$. The residuals are:

rc	$-r$	$-r$	$-r$	$-r$	$\cdots$	$-r$
$-c$	1	1	1	1	$\cdots$	1
$-c$	1	1	1	1	$\cdots$	1
$-c$	1	1	1	1	$\cdots$	1
.	.	.	.	.	$\cdots$	1
$-c$	1	1	1	1	$\cdots$	1

Yates also wrote that "when a set of interactions is found to be significant, there is a probability that the whole of this significance may be accounted for by a single outstanding value." He tabulates the contrast $D - (A + B + C)/3$ for each center and judges that it is notably larger for center 4. The one-celled interaction must have been spotted before construction of this contrast. Our usual rigmarole for estimating residuals is repeated in Table 8.1. In its lower panel, the table gives the pattern of attenuation of a single exceptional disturbance in a general $R \times C$ table.

It is evident from Table 8.1, b, that a maximum residual of size d_{ij} implies an estimated disturbance of $(RC/rc)\, d_{ij}$. For the data of Table 8.1, panel a, we see that $(28/18)12^2 = 224$ is removed from the original 335. The remainder of 111 with 17 d.f. gives a decoded residual MS of 0.000658, less than the MS for error given by Yates of 0.000869. We have again found a single aberrant cell which accounts for *all* of the visible interaction.

8.3. A $5 \times 3 \times 4$ FROM DAVIES [1971, ED. 2, PAGES 291–295]

This was a three-factor unreplicated factorial design, all factors (A, B, C) at discrete levels. The 5×4 table for $A \times C$ (Table 8.2) is chosen to make

TABLE 8.2.

$A \times C$ INTERACTION FROM DAVIES' $5 \times 3 \times 4$ [1971, PAGE 294], DATA CODED BY -954 AND ROUNDED TO 10's

Panel a

		Coded Data						Residuals			
		C_1	C_2	C_3	C_4	Sums	Averages				
	A_1	46	39	13	51	149	37	-3	10	-3	-3
	A_2	29	-21	-36	33	5	1	16	-14	-16	15
	A_3	24	23	0	31	78	19	-7	12	2	-5
	A_4	-3	-31	-29	-1	-64	-16	1	-7	8	-2
	A_5	-34	-47	-49	-28	-158	-39	-7	0	11	-6
Sums:		62	-37	-101	86	10		SS $(AC) = 1606/3 = 535.$			
Sums $-$ 2:		60	-39	-103	84	2		MS $(AC) = 535/12 = 45.$			
Deviations:		12	-8	-21	17						

Panel b. With row A_2 removed

	Residuals			
	1	6	-7	1
SS $[(AC) - A_2] = 406/3 = 135.$	-4	7	-3	1
MS $\quad= 135/9 = 15.$	5	-11	4	2
Decoded $\quad= 1500.$	-3	-4	7	-2

a petty point. The four largest residuals are in row A_2 (Filler Quality). We do need a test for residual heterogeneity for doubtful cases, but this is hardly a doubtful case. Filler Quality 2 gives wear resistances with the four qualities of rubber (factor C) in a different pattern from the other four fillers. We set row 2 aside and repeat the computation in the lower part of Table 8.2. We see that there is still some $A \times C$ interaction when compared to the original MS (ABC) of 320 (3.20 in our coding). We have located only two thirds of the AC interaction in row A_2.

Contrary to the remark in Davies ("In the general case, the interaction sum of squares cannot be conveniently calculated by direct methods and it is usual to derive it by subtraction from the total" [1971, page 294]), we find it both convenient and more illuminating to compute the interaction cell by cell and to get the interaction SS directly by squaring and summing the residuals. The reader—used to desk or computer results to many more significant figures than the work shown—may be uncomfortable at the sight of such gross rounding. He should be reminded that the high precision to which he is accustomed is required only when some SS's are to be estimated by subtraction. The method shown does not get any SS in that way.

It is appropriate to show the computation of the three-factor interaction, ABC, here, even though it does not strictly fall under the title of this chapter. The ABC interaction may be viewed as $(AB)C$, as $(AC)B$, or as $(BC)A$. Since B has the smallest number of levels, the arithmetic is lightened a little by choosing the second alternative. This requires that we make up an $A \times C$ table for each level of B, and find the three sets of residuals. The deviations of these from their averages, cell by cell, give the 60 components of the three-factor interaction. In Table 8.3 these are put into the same arrangement as the original data. Nothing remarkable emerges, unless we clutch at the straw floating by in the form of the three largest components (-27, -25, and 25), all in row A_4. The table is given so that the reader can, perhaps for the first time, see all the terms of a three-factor interaction.

TABLE 8.3.
COMPONENTS OF THE ABC INTERACTION FOR THE $5 \times 3 \times 4$ OF
DAVIES [1971, PAGE 291]

	C_1			C_2			C_3			C_4		
	B_1	B_2	B_3	B_1	B_2	B_3	B_1	B_2	B_3	B_1	B_2	B_3
A_1	-10	10	-1	3	-15	13	18	-6	-13	-10	12	0
A_2	-12	4	4	14	4	-19	6	-13	3	-11	2	8
A_3	-4	11	0	-15	0	13	19	-4	-17	1	-7	6
A_4	19	-27	7	-7	7	1	-25	25	0	14	-5	-6
A_5	8	3	-11	3	3	-6	-18	-6	22	9	-3	-5

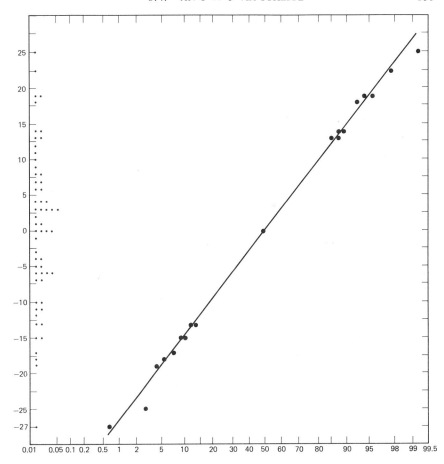

Figure 8.1 E.c.d. for the 60 components of $A \times B \times C$ from Davies' $5 \times 3 \times 4$. $\hat{\sigma} = 11.5\sqrt{59/24} = 18$. See Table 8.3.

Figure 8.1 shows, on a normal grid, the outer 16 terms plus the 30th term. The observed s of 11.5 is of course the root mean square of 60 values, but since 36 constants have been fitted, $(1 + 4 + 2 + 3 + 8 + 12 + 6 = 36$ d.f. for mean, A, B, C, AB, AC, BC, respectively) only 24 d.f. remain. We make a rough correction by multiplying by $(59/24)^{1/2} = 1.569$ to get 18.05, which matches nicely the 17.9 given by Davies.

8.4. AN 8 × 5 VIA SCHEFFÉ [1959, HIS TABLE B, PAGE 138]

The 8 × 5 data set from a randomized block experiment on eight varieties of oats, originally from Anderson and Bancroft [1952], is large enough to provide some evidence on the shape of the error distribution. The residuals, shown in panel *b* of Table 8.4, are plotted in Figure 8.2 on an "arithmetic normal grid." The plotting is done in two stages. First the grid is scaled, in this case from -80 to $+80$. Then each residual is picked up from the residual table and plotted as a dot near the left margin, as shown in the figure. In this way, we avoid the necessity of searching for the largest, then for the next largest, etc. When all are transferred, we count them to make sure that none has been lost.

Each point is now moved in to a per cent probability point, which is

$$P' = 100\,\frac{i - \frac{1}{2}}{N}$$

for the *i*th point.*

The per cents can be read from a slide rule with ample precision, and two points (one at each end of the cumulative distribution) can be plotted for each per cent computed. The straight line is drawn by eye, using a transparent rule, and is forced through the (0, 50%) point. The data near the 16 and 84% points should be given somewhat greater (vertical) weight in deciding on the slope of the line. The error std. dev. is most easily estimated as *half* the difference on the residual scale between the 16% and 84% points taken from the line drawn. Since the $N = RC$ residuals have only $(R - 1)(C - 1)$ degrees of freedom, we get a fairer estimate of the standard deviation of observations by multiplying the graphical value by $[RC/(R - 1)(C - 1)]^{1/2}$.

There is a crudity in this correction that may be deplored, but that does not seem to me to be serious for $R \times C$ tables larger than 5×5. It is well

---

* I am aware of some differing opinions on the choice of plotting positions—in particular, those of Blom [1958], Harter [1969], and Tukey [1962]. Blom recommends for α in the equation

$$P'' = 100\,\frac{i - \alpha}{N + 1 - \alpha}$$

the value $\frac{3}{8}$ as a general compromise, since the optimum value varies with N. Harter shows that this value (0.375) is generally a bit low, but for $N = 20$ he gives 0.378 and for $N = 50, 0.389$.
If we use, then,

$$P'' = \frac{i - 0.38}{N + 0.62},$$

we have for $N = 40$

i	P'	P''
1	1.250	1.526
2	3.75	3.99
3	6.25	6.44

The discrepancy for $i = 1$ is less than $\frac{1}{8}$ inch on the usual $8\frac{1}{2} \times 11$ inch grid and decreases for larger i. Deviations of this magnitude are negligible compared to those that we will be judging important.

TABLE 8.4.

AN 8 × 5 FROM ANDERSON AND BANCROFT, VIA SCHEFFÉ [1959, PAGE 138], DATA CODED BY −354

Panel a								Panel b				
	Blocks							Residuals				
Variety	I	II	III	IV	V	$\sum_R \div 5$		I	II	III	IV	V
1	−58	3	−14	−23	−6	−98	−20	−67	20	−23	10	62
2	48	36	77	−14	−34	113	23	−4	10	25	−24	−9
3	83	−20	72	−34	−58	43	9	45	−32	34	−30	−21
4	−51	−35	−44	−94	−112	−336	−67	−13	29	−6	−14	3
5	115	51	88	133	40	427	85	1	−37	−26	61	3
6	−9	−12	4	−54	−46	−117	−23	−15	8	−2	−18	25
7	−30	−15	3	−2	−134	−178	−36	−23	18	10	47	−50
8	134	20	47	−16	−34	151	30	75	−13	−12	−33	−16

$\sum_C$:	232	28	233	−104	−384	5	Residual MS_1	= 37,514/28
$\sum_C − 1$:	231	27	232	−105	−385	0		1340.
$\sum_C \div 8$:	29	3	29	−13	−48		$s_1 = 36.6$.	
							See Figure 8.2.	

Panel c						Panel d					
	Residuals$_2$, Block I Removed						Residuals$_1$, Ordered by $\hat{\alpha}_i$ and $\hat{\beta}_j$				
Variety	II	III	IV	V		Variety	III	I	II	IV	V
1	3	−42	−7	46		5	−26	1	−37	61	3
2	10	23	−24	−8		8	−12	75	−13	−33	−16
3	−20	44	−18	−6		2	25	−4	10	−24	−9
4	26	−11	−17	1		3	34	45	−32	−30	−21
5	−39	−16	59	−2							
6	5	−7	−21	23		1	−23	−67	20	10	62
7	12	2	41	−55		6	−2	−15	8	−18	25
8	6	5	−14	4		7	10	−23	18	47	−50
Residual MS_2 = 20,423/21 = 972.5.						4	−6	−13	29	−14	3
$s_2 = 31, 2$											

known that the correlation coefficient between pairs of residuals in the same row (column) is $−1/(C − 1) \left[−1/(R − 1)\right]$, while that between residuals in different rows and columns is $−1/(R − 1)(C − 1)$. This disparity becomes less when R and C are larger.

Panel b of Table 8.4 shows the residuals from the usual additive model. This suggests, without being entirely convincing, that block I was more disturbed than the other four. Carrying through the analogous arithmetic for blocks II, III, IV, and V, we acquire the residuals and summary statistics

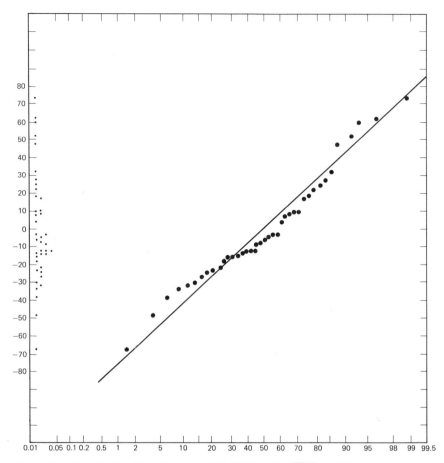

Figure 8.2   E.c.d. for the 40 residuals from Scheffe'. $\hat{\sigma} = 33\sqrt{39/28} = 39$. See Table 8.4.

shown in panel c. The normal plot of the new residuals looks "better," but this opinion is subjective and tendentious.

It is natural to consider applying Tukey's G-test (Section 4.6), but a simple rearrangement of the table shows this to be unnecessary. Panel d gives the 8×5 display of residuals, but rearranged by (decreasing) row and column averages. As Scheffé mentions [1959, page 132], Tukey's statistic can be written as

$$G = \frac{\sum_i \sum_j \hat{\alpha}_i \hat{\beta}_j \hat{\gamma}_{ij}}{\sum_i \hat{\alpha}_i^2 \cdot \sum_j \hat{\beta}_j^2},$$

The rearrangement of panel c puts the $\hat{\alpha}_i$ and $\hat{\beta}_j$ into decreasing orders and so induces maximum positive correlation between them as they stand. If G is to be large, the $\hat{\gamma}_{ij}$ must be in the general form of a "linear by linear" interaction contrast. This means a predominance of residuals of the same sign in the diagonal quadrants with the largest residuals near the corners. Since Panel d does not reveal this pattern, the detailed test is not carried through.

I conclude that these data are satisfactory in the sense that the standard assumptions (normal, uncorrelated, constant-variance observations, with additive row and column effects) are satisfied for blocks II–V. There appear to be two exceptional values (for varieties 1 and 8) in block I.

8.5.   A 6 × 5 × 2 ON BARLEY (IMMER ET AL.) FROM FISHER [1953]

The data are given as a "practical example" by R. A. Fisher [1953, page 66]. Results are shown for the total yields of five varieties of barley in two successive years at six locations in the state of Minnesota. We can view the data table as two 6×5 layouts, one summed over the two years, the other showing the difference between the 2 years. Although the data were originally given to one decimal place, they are rounded to the nearest unit in Table 8.5, panel a. Also, 101, the approximate mean, has been subtracted from each value.

Panels b and c of Table 8.5 show the sums over and the differences between the yields in the two years. We look first at the differences and at the corresponding table of residuals. The large residual in row 5, at column 3, is .05 significant by the maximum normed residual test. (These residuals are of course just twice the components, $\hat{\delta}_{ijk}$, of the three-factor interaction.) Inspection of the corresponding difference in panel c shows it to be -28, while the other items in the same row are 32, 39, 27, and 24. The sum of the corresponding two items (in panel b) is 36, close to the average of the other entries in that row. These facts suggest that the two entries have been interchanged in error. If this is so, the error MS should be revised from its original value of 143 to 89. This has serious consequences for judgments on main effects and on two-factor interactions. The only other serious change would be in the overall yearly difference, which is not now (and was not then) an important parameter.

Turning now to panel b, we can see that all of the four largest $\hat{\gamma}_{ij}$ are for variety 4 (Trebi), which gave the highest average yield. The remaining MS for six locations by four varieties is 109, a value negligibly larger than the corrected error MS of 89 (3fi).

TABLE 8.5.

YIELDS OF BARLEY IN SIX LOCATIONS, FOR FIVE VARIETIES, IN TWO YEARS; DATA BY IMMER ET AL. FROM FISHER [1953], ROUNDED TO 1 AND CODED BY −101

Panel a.

Location	Year	M	S	V	T	P
1	1	−20	4	19	9	−3
	2	−20	−19	−21	−14	−17
2	1	46	41	50	91	45
	2	−1	15	11	47	7
3	1	−19	−24	−23	30	−11
	2	2	4	16	39	29
4	1	19	20	23	40	24
	2	−2	−39	−5	25	−25
5	1	−2	−12	−32	−12	3
	2	−35	−51	−4	−39	−21
6	1	−14	−24	−22	1	−5
	2	−33	−34	−34	−9	−7

Panel b.

Sums over Years

Locations	M	S	V	T	P	Averages	Residuals				
1	−40	−15	−2	−5	−20	−16	−11	21	18	⟨−24⟩	−7
2	45	56	61	138	52	70	−12	6	−5	⟨33⟩	−21
3	−17	−20	−7	69	18	9	−13	−9	−12	⟨25⟩	6
4	17	−19	18	65	−1	16	14	−15	6	14	−20
5	−37	−63	−36	−51	−18	−41	17	2	9	⟨−45⟩	20
6	−47	−58	−56	−8	−12	−36	2	−2	−16	−7	21

Column deviations: −13 −20 −4 35 3

Panel c.

Locations	Year Differences (1931–1932)					Averages	Residuals				
1	0	23	39	23	14	20	−20	−3	26	1	−5
2	46	26	39	44	38	39	7	−19	7	3	0
3	−21	−28	−38	−9	−40	−27	6	−7	−4	16	−12
4	21	59	28	15	49	35	−14	18	0	−22	15
5	32	39	−28	27	24	19	13	14	⟨−40⟩	6	6
6	19	10	12	10	2	11	8	−7	8	−3	−8

Column deviations: 0 6 −7 2 −1

It is natural to try Tukey's test for multiplicative nonadditivity (called the G-test below), on these data. The details are given in Table 8.6 because they reveal that, even when a test turns out to be significant, there may be reasons for not accepting the outcome at face value. We reorder the table of residuals ($\hat{\gamma}_{ij}$) of panel a, Table 8.5, by row and column means to get panel a of Table 8.6. We do see the largest positive residual at cell (1, 1) and the largest negative one at cell (5, 1), but the agreement is not striking elsewhere. Since the F-test for G is significant, we proceed to compute $\tilde{G}$ and the resulting residuals, $\tilde{\gamma}_{ij} = \tilde{G}\hat{\alpha}_i\hat{\beta}_j$. These are shown in panel b. The resemblance of these doubly

TABLE 8.6.

TUKEY'S G-TEST ON RESIDUALS OF TABLE 8.5, PANEL b, REORDERED BY ROW AND COLUMN MEANS

Panel a		Varieties						
Locations	4	5	3	1	2	Averages	$\hat{\alpha}_i$	p_i
2	33	−21	−5	−12	6	136	35	553
4	14	−21	6	14	−15	109	8	236
3	25	6	−12	−13	−9	105	4	642
1	−24	−7	18	−11	21	93	−8	−591
6	−45	20	9	17	−2	83	−18	−842
5	−7	21	−16	2	−2	80	−21	−39
Deviations ($\hat{\beta}_j$):	17	2	−2	−7	−10			

$p_i = \sum_j y_{ij}\hat{\beta}_j$; $P = \sum_i \alpha_i p_i = 44{,}514$.

$\sum \hat{\alpha}_i^2 = 2134$; $\sum \hat{\beta}_j^2 = 446$.

SS (nonadditivity) $= 44{,}514^2/(2134 \times 446) = 2082$.

F (nonadditivity) $= 2082/(2 \times 143) = 7.28$; $P_F < .025$.

$\tilde{G} = 44{,}514/(2134 \times 446) = 0.0468$.

Panel b									
Residuals Predicted by $\tilde{G}$ $\tilde{\gamma}_{ij} = \tilde{G}\hat{\alpha}_i\hat{\beta}_j$					Residual Residuals $= \hat{\gamma}_{ij} - \tilde{\gamma}_{ij}$				
28	3	−3	−11	−16	5	−24	−2	−1	10
6	1	−1	−2	−4	8	−22	7	16	−11
3	0	0	−1	−1	22	6	−12	−12	−8
−6	−1	1	2	4	−18	−6	17	−13	17
−14	−2	2	6	8	−31	22	7	11	−10
−17	−2	2	7	10	10	23	−18	−5	−12

ordered residuals to a "linear by linear" interaction contrast is strong. Subtracting the "predicted residuals" of panel b from those of panel a, we reach the "residual residuals" of panel c. Some of the original residuals have been nicely reduced, but others have not changed and a third group has even been increased. It is obvious that nearly all of the reduction in residual SS produced by the G-transform has come from the two variety 4 (now in column 1) residuals $+33$ and -45, spotted earlier. It is also noteworthy (and deplorable) that all the residuals for location 5 (last row) have been increased.

We must decline, then, to give strong weight to the significance of the G-test in this case, since its value depends so largely on two residuals.

Locations 5 seems to have produced less reliable data than the other locations. Two of its values (for variety 3) appear to have been transposed; the largest residual in Table 8.5, panel b (for Trebi), is to be laid at its door, and it had the lowest yield overall. We have therefore dropped the data from this location and redone the whole analysis, but, as the weary reader will be relieved to see, we do not display all the arithmetic. There remain consistent differences between varieties, locations, and even years. The yearly differences were consistent for four locations but were reversed for location 2. No large location-variety interaction remains. The error std. dev. (3fi) is reduced to 10 as compared to the 12 computed from the full set of data.

8.6. A 33 × 31 FROM PFISTER [LIGHTFASTNESS (y) OF 33 DYE BASES (A), EACH COUPLED TO 31 NAPHTHOLS (B)]

The data were published by Pfister Chemical Company [1955] and were given as scores from 1 to 8, that is, from "very poor" to "excellent," with occasional $+$ or $-$ suffixes to indicate intermediate scores. These have been dropped so that only integers appear in Table 8.7. It is not plausible to assume that the random error in these observations is normally distributed with constant variance, nor is it likely that the observations were taken under randomized or other statistically independent conditions. The code letters at the head of each column identifying the naphthols in the Pfister publication are commercial designations, related to chemical composition. All codes containing the letter G are chemically related, and those containing LG are a more closely related subset.

Since both row and column categories are discrete, not ordered or continuous, any rearrangement of rows or column loses no information. Indeed we will see that rearranging by decreasing row and column averages produces considerable clarification and eases computation. Table 8.8 shows the rearranged data. The expected and visible result of this reordering is to surround most observations with numbers of similar magnitudes. Exceptions to this near matching are then evidences of nonadditivity. The ordered table

TABLE 8.7.

Lightfastness of 33 Dye Bases × 31 Naphthols; Data from Pfister Chemical Company [1955]

1 = Very Poor....8 = Exceptional*

Dye Bases	1 O	2 AN	3 BG	4 BO	5 BR	6 BS	7 CL	8 D	9 DB	10 E	11 G	12 GR	13 TTR	14 KB	15 LB	16 LC	17 LG	18 L3G	19 L4G	20 LT	21 MCA	22 MX	23 OL	24 P	25 RL	26 RP	27 RT	28 SG	29 SR	30 SW	31 TR	∑_R	Row Rank
1	5	4	5	5	4	4	6	5	5	4	5	3	5	5	6	5	5	4	6	6	4	6	4	6	6	4	5	4	7	5	5	155	15
2	5	4	6	5	4	4	6	5	7	5	6	3	5	6	6	6	5	4	6	6	3	6	5	6	6	6	5	4	7	5	6	159	12
3	5	4	7	6	4	4	6	6	6	5	4	4	6	6	6	7	5	5	1	7	5	6	6	6	7	5	6	3	7	5	6	163	9
4	5	3	6	5	4	3	5	6	6	3	4	3	6	4	6	6	5	5	6	6	3	4	7	5	6	6	6	4	7	5	4	138	28
5	6	4	6	6	4	4	6	6	6	5	5	4	7	6	6	6	5	6	4	6	4	5	7	7	6	6	4	3	6	4	6	174	3
6	4	4	6	6	5	4	6	6	6	3	1	3	6	5	6	6	5	4	7	6	5	6	6	6	5	6	4	7	7	4	6	149	20
7	5	4	6	7	5	4	6	5	6	3	3	3	6	5	6	6	5	7	7	6	3	5	5	6	5	6	5	7	7	6	5	179	2
8	5	1	5	5	5	4	6	5	6	3	4	3	6	5	7	6	5	4	4	6	4	6	5	6	4	5	6	7	7	5	5	147	21
9	4	4	5	5	4	4	6	5	6	3	4	5	7	6	7	5	5	5	5	7	5	6	7	7	7	6	6	7	6	7	5	151	18
10	6	6	7	7	6	3	6	7	7	4	4	4	6	6	7	7	5	3	6	6	4	7	6	7	7	6	6	6	6	5	6	193	1
11	5	6	5	6	5	4	6	5	6	5	4	5	6	5	7	7	5	5	6	7	5	6	5	5	7	6	5	7	6	7	6	165	6
12	6	2	6	6	4	4	6	5	6	3	4	4	6	5	7	7	4	3	5	5	4	6	7	7	7	6	5	6	6	5	6	161	11
13	5	4	5	6	5	4	6	7	7	5	4	4	6	4	6	6	4	5	5	6	5	5	6	5	6	6	5	6	7	6	4	163	10
14	6	4	6	6	4	4	7	5	6	4	4	4	6	6	6	7	5	5	5	7	3	5	7	6	7	6	5	6	7	6	5	169	4
15	3	3	5	5	3	3	4	6	6	5	4	3	7	5	7	4	4	6	6	5	4	5	4	6	5	4	5	6	6	3	5	142	25
16	4	3	6	6	5	4	6	5	6	3	5	4	6	5	6	6	5	6	6	6	4	6	6	5	6	6	5	6	6	5	5	159	13
17	4	4	6	6	4	4	6	5	6	5	5	3	7	5	7	6	5	6	5	6	5	6	6	6	6	5	5	6	6	5	4	168	5
18	5	4	6	6	5	4	6	5	7	2	4	3	6	6	7	6	4	2	5	7	2	4	6	4	7	6	5	6	7	4	6	153	17
19	4	4	6	6	4	4	6	5	6	3	4	2	6	5	7	6	4	3	5	7	4	5	6	6	7	6	5	6	7	4	6	165	7
20	4	6	4	4	4	3	6	6	5	3	4	3	6	5	7	6	4	3	4	6	4	4	6	5	4	6	5	4	7	4	6	145	24
21	5	4	5	4	4	4	6	4	4	4	2	2	6	5	5	5	2	2	4	6	4	5	4	6	5	6	5	4	6	4	5	151	19
22	4	5	5	5	4	4	5	6	4	3	3	2	5	4	5	6	2	2	4	5	3	4	4	7	5	6	5	4	5	4	4	124	31
23	5	4	5	6	2	5	5	4	4	4	4	4	7	6	6	5	1	1	4	5	4	5	7	6	6	6	4	6	7	5	5	139	27
24	6	4	6	5	5	5	5	4	7	4	4	2	6	4	7	6	7	3	6	6	4	6	4	5	6	6	4	6	7	5	4	165	8
25	5	4	5	6	5	5	5	5	2	4	3	4	7	6	7	6	3	3	5	5	4	6	6	5	6	6	4	6	6	6	6	158	14
26	5	4	4	5	3	5	4	4	6	2	3	2	4	3	6	3	2	2	3	3	2	4	5	5	4	5	4	4	5	4	3	127	30
27	5	4	5	3	2	4	4	5	6	2	2	2	4	4	6	4	1	1	1	3	5	4	5	5	5	5	5	5	5	5	5	118	32
28	5	3	5	5	2	5	6	5	7	5	2	2	5	5	6	4	1	1	4	6	5	5	4	5	5	5	5	4	5	4	3	136	26
29	6	6	5	5	2	6	6	6	5	5	2	6	6	4	4	4	1	1	4	6	5	6	6	5	5	5	6	6	6	5	5	146	22
30	6	6	5	5	3	5	5	6	7	3	1	4	3	6	6	3	1	1	4	3	5	5	6	5	3	3	6	5	6	5	5	148	23
31	3	6	4	3	3	6	5	3	1	4	6	4	6	4	4	4	1	1	4	6	4	6	3	5	5	3	6	4	5	4	4	103	33
32	5	5	3	3	3	5	5	6	2	3	1	4	3	5	6	3	1	1	2	5	4	5	5	5	5	5	3	5	5	5	5	138	29
33	5	6	6	6	5	6	6	6	4	4	6	4	5	6	6	5	2	2	2	4	4	5	5	5	5	5	6	5	6	6	6	155	16
∑_C:	160	132	179	172	133	141	175	168	179	120	124	114	186	165	200	177	122	120	144	192	131	179	180	183	181	177	167	176	197	163	168	5014	
Column rank:	21	25	8	15	24	23	14	16	9	27	28	31	4	19	1	11	29	30	22	3	26	10	7	5	6	12	18	13	2	20	17		

* Rounded down.

TABLE 8.8.
Dye Bases and Naphthols Reordered by Average Lightfastness

Rank	Dye Base	1 LB 15	2 SR 29	3 LT 20	4 ITR 13	5 P 24	6 RL 25	7 OL 23	8 BG 3	9 DB 9	10 MX 22	11 LC 16	12 RP 26	13 SG 28	14 CL 7	15 BC 4	16 D 8	17 TR 31	18 RT 27	19 KB 14	20 SW 30	21 O 1	22 L4G 19	23 BS 6	24 BR 5	25 AN 2	26 MCA 21	27 E 10	28 G 11	29 LG 17	30 L3G 18	31 GR 12	Averages
1	10	5	6	6	7	7	7	7	7	6	7	7	6	6	7	6	6	7	6	6	7	6	6	6	6	6	5	5	5	6	7	5	6.2
2	7	6	7	6	6	6	6	6	7	6	6	6	6	7	6	6	6	5	6	5	6	5	7	4	5	4	5	5	7	6	7	3	5.8
3	5	6	6	6	6	6	6	6	6	6	7	6	6	3	6	5	5	6	6	6	5	6	7	4	5	4	5	4	5	6	6	4	5.6
4	14	6	5	7	6	6	6	7	6	4	5	6	6	5	7	6	5	6	5	6	6	6	5	4	5	4	5	4	5	5	6	4	5.4
5	17	7	6	6	6	6	7	6	6	6	6	7	6	6	5	6	5	5	5	5	5	4	6	4	5	2	5	5	5	5	6	4	5.4
6	11	7	7	7	6	5	7	6	6	7	5	6	6	7	6	6	5	6	5	6	4	5	5	4	5	4	4	4	4	4	3	4	5.3
7	19	7	7	6	7	4	6	7	6	7	4	6	6	6	6	5	5	4	5	4	5	5	6	4	5	4	4	4	4	4	5	4	5.3
8	24	7	7	7	5	6	7	7	6	7	6	6	6	6	6	5	5	5	4	5	5	5	1	4	4	4	5	4	4	7	5	4	5.3
9	3	6	7	6	6	6	7	6	7	6	6	6	6	6	6	5	4	6	6	6	5	5	5	4	4	4	5	4	4	4	5	4	5.3
10	13	6	6	5	5	6	6	6	6	6	5	6	6	3	7	6	6	4	5	4	6	6	5	3	4	3	5	5	4	5	4	3	5.3
11	12	7	6	6	5	5	6	6	6	7	5	6	6	6	6	5	5	6	5	6	5	5	6	5	4	4	3	5	4	4	4	3	5.2
12	2	6	7	5	7	6	6	6	6	7	5	6	6	4	6	5	5	5	5	6	5	5	6	3	4	3	4	3	6	4	4	3	5.1
13	16	7	6	5	6	5	6	5	5	6	6	6	6	6	4	5	5	5	5	5	5	4	6	3	4	4	4	4	5	5	6	3	5.1
14	25	6	6	6	5	6	6	5	6	2	5	5	4	5	6	5	5	5	5	5	5	6	6	5	5	6	4	4	6	3	6	4	5.0
15	1	6	6	5	5	6	6	5	5	4	6	5	5	4	6	6	5	5	5	6	5	6	6	3	5	4	4	4	5	5	6	3	5.0
16	33	7	6	6	6	6	6	6	4	6	6	5	5	5	6	6	6	5	5	5	5	5	2	6	5	6	4	4	4	2	2	3	4.9
17	18	7	7	6	6	6	6	6	5	6	6	5	5	6	5	5	4	4	5	4	5	5	5	4	4	3	3	2	5	4	2	3	4.9
18	9	7	6	5	7	6	6	7	5	5	6	5	5	6	6	5	5	5	5	5	5	4	4	5	5	3	5	4	4	5	2	3	4.9
19	21	5	5	6	5	6	6	6	5	6	6	5	6	6	5	6	4	5	4	5	5	6	5	3	5	3	3	4	4	4	2	2	4.9
20	6	6	6	6	6	6	6	6	5	6	5	6	5	4	5	6	4	5	5	5	5	5	4	4	5	4	4	3	4	5	2	3	4.8
21	8	6	6	5	6	6	5	5	5	6	5	6	5	7	5	5	5	5	5	4	5	5	4	4	2	1	3	3	2	2	4	2	4.7
22	29	6	6	6	5	5	4	5	5	1	6	5	5	7	5	5	5	5	6	5	5	5	4	4	5	6	5	5	2	1	2	3	4.7
23	30	6	5	5	6	6	5	6	5	6	6	5	6	6	6	5	5	5	5	5	5	6	4	5	2	4	3	3	2	4	1	3	4.7
24	20	7	6	6	6	6	6	5	4	6	5	4	5	5	4	4	5	6	6	4	4	5	6	6	3	3	3	3	4	4	3	3	4.7
25	15	6	6	5	5	6	6	5	5	6	6	5	5	6	6	5	5	5	5	5	3	4	3	3	3	3	3	3	2	1	5	3	4.6
26	28	6	6	6	5	6	5	5	5	4	5	5	5	4	5	5	5	5	4	5	5	5	4	5	2	3	5	5	4	5	1	4	4.6
27	23	5	7	6	5	5	5	5	4	6	6	5	5	4	5	5	5	4	5	4	5	4	4	3	2	3	5	3	2	1	2	2	4.5
28	4	6	5	5	6	6	5	5	5	4	5	5	5	5	5	6	5	5	4	4	4	4	1	3	4	5	4	4	2	1	4	2	4.5
29	32	6	5	6	5	5	5	4	5	6	5	5	4	4	5	4	4	5	5	5	5	5	3	3	3	3	4	4	2	4	2	4	4.4
30	26	6	5	5	4	6	5	4	4	4	4	5	5	4	6	5	5	4	4	4	4	5	3	4	3	4	2	3	2	1	2	2	4.1
31	22	6	5	3	5	5	5	4	5	6	4	4	5	4	4	5	4	5	5	3	5	4	1	4	3	3	3	2	2	2	2	2	4.0
32	27	6	5	3	4	5	5	5	4	4	4	3	3	4	4	5	5	3	4	4	5	4	1	3	3	4	2	2	4	1	2	2	3.8
33	31	4	5	3	3	5	3	3	4	2	4	3	3	5	4	3	5	4	5	4	4	3	1	5	3	5	3	3	1	1	1	4	3.3
Column Deviations:		1.2	1.1	1.0	.7	.6	.6	.6	.5	.5	.5	.5	.5	.5	.4	.3	.2	.2	.1	.1	0	-.1	-.5	-.6	-.9	-.9	-.9	-1.0	-1.2	-1.2	-1.3	-1.4	4.9

is bordered with row averages and column deviations so that fitted values
and residuals can be computed without pencil. For example, the average
for row 1 is 6.2 and the column deviation for column 1 is 1.2, so the fitted
value in cell (1, 1) is 7.4. The residual is then (5 − 7.4) or −2.4. Is this large
or small? We can get a (conservative) view of this by evaluating the residual
root mean square for the whole table, including, at first, all interactions.

The "analysis of variance identity" for a two-way table (which holds without any assumption about the distribution of error) can be used to partition
each observation into *four* components: a grand mean, a row "effect" or
deviation from the grand mean, a column effect or deviation from the grand
mean, and a residual term. Using $y_{..}$ to denote the grand average, $y_{i.}$ for the
*i*th row average, and $y_{.j}$ for the *j*th column average, we can write the identity
as

(8.1) $y_{ij} = y_{..} + (y_{i.} - y_{..}) + (y_{.j} - y_{..}) + (y_{ij} - y_{i.} - y_{.j} + y_{..})$

(8.2)   or $= y_{i.} + (y_{.j} - y_{..}) + (y_{ij} - y_{i.} - y_{.y} + y_{..})$

(8.3)   or   = Row average + Column deviation + Residual

(8.4)   or $= A_i + \hat{\beta}_j + \hat{\gamma}_{ij}.$

From (8.4) it is easy to compute the $\hat{\gamma}_{ij}$. By squaring both sides of (8.1) and
summing over *i* and *j*, we obtain the corresponding sum of squares identity;
then, transposing to isolate the SS for interaction, we have

(8.5) $\sum_i \sum_j \hat{\gamma}_{ij}^2 = \sum_i \sum_j (y_{ij} - y_{..})^2 - C \sum_i (y_{i.} - y_{..})^2 - R \sum_j (y_{.j} - y_{..})^2.$

since all cross-product terms vanish identically. The three sums on the right
are easily computed. All are given in Table 8.10 in standard analysis of
variance form.

The square root of the residual MS estimates the error std. dev. if there
are no interactions. This is hardly likely, however, so we have in $0.82^{1/2} =$
0.91 a somewhat inflated estimate. We have gone carefully through ordered
Table 8.8, spotting discrepant values by their failure to nearly match their
neighbors, estimating their residuals, and circling them (in Table 8.9) if the
latter are as large as 2.1. There appear to be 31 such entries. Figure 8.3 shows
these residuals plotted as circled dots on a normal grid. A straight line through
the upper six points *and* through the (50%, 0) point suggests that quite a few
of the negative residuals are not part of the implied normal distribution.
Indeed, if *six* of these, all less than −3, are removed, the remaining circled
points find themselves moved over onto the straight line (as x's). I suppose,
then, that we have a nearly normal distribution of error holding for all but
six of our 1023 points. The empirical cumulative distribution found suffices

TABLE 8.9.

Dye Bases and Naphthols Reordered by Average Lightfastness; Discrepant Values Circled

Six largest discrepancies designated by asterisk.

P	1	2	3	4	5	6	7	8	9	10	11	12	13	14	15	16	17	18	19	20	21	22	23	24	25	26	27	28	29	30	31		
	LB	SR	LT	ITR	AN	RL	OL	BG	DB	MX	LC	RP	SG	CL	BO	D	TR	RT	KB	SW	O	L4G	BS	BR	AN	MCA	E	G	LG	L3G	GR		
P Rank / Dye Base	15	29	20	13	24	25	23	3	9	22	16	26	28	7	4	8	31	27	14	30	1	19	6	5	2	21	10	11	17	18	12	Averages	n_i
1 / 10	(5)	6	6	6	7	7	7	7	6	7	7	6	6	7	6	6	7	6	6	7	6	7	6	6	6	5	5	5	6	7	5	6.2	1
2 / 7	6	7	6	7	7	6	7	7	6	7	6	6	7*	6	6	6	6	6	5	6	5	7	6	6	4	5	5	(7)	6	(7)	5	5.8	2
3 / 5	6	6	6	7	6	6	7	6	6	7	6	6	(3)*	6	5	5	6	6	6	6	6	7	5	5	4	4	4	5	5	6	3	5.6	1
4 / 14	6	6	7	6	6	6	7	6	4	5	7	6	6	7	5	5	5	5	6	6	6	5	4	5	(2)	5	5	4	5	6	4	5.4	
5 / 17	6	6	7	6	6	6	6	6	7	5	7	6	6	7	6	5	4	5	4	5	5	5	4	5	4	5	5	4	5	6	4	5.4	1
6 / 11	7	7	7	6	5	7	6	6	7	4	6	6	6	5	5	5	4	4	4	5	5	6	4	5	4	4	4	4	4	4	4	5.3	
7 / 19	7	7	7	6	4	6	6	6	7	4	6	6	6	6	6	5	4	5	4	5	5	6	4	5	4	5	5	4	4	5	4	5.3	1
8 / 24	7	5	7	7	6	7	6	5	6	6	6	6	5	6	5	6	6	6	4	4	5	5	4	4	4	4	4	4	(7)*	5	4	5.3	2
9 / 3	6	6	7	6	6	6	6	6	7	6	6	6	6	6	6	5	5	5	5	5	5	5	4	4	4	5	5	4	5	5	4	5.3	
10 / 13	6	6	6	6	5	5	6	6	6	5	6	6	6	7	6	6	6	5	4	5	5	6	4	4	4	5	5	4	4	5	4	5.3	
11 / 12	6	6	6	5	5	5	5	5	6	5	6	6	6	5	5	5	5	5	4	5	5	(1)*	4	4	4	5	5	4	4	5	3	5.2	1
12 / 2	5	6	6	5	5	5	5	6	(2)*	5	5	4	6	6	5	5	6	5	6	5	5	5	3	4	4	3	3	(6)*	4	(6)*	3	5.1	1
13 / 16	5	6	6	5	5	6	6	6	5	5	5	5	6	5	5	5	6	5	5	5	4	6	5	4	3	4	4	5	3	3	4	5.1	1
14 / 25	5	6	6	5	5	6	6	6	4	6	5	5	6	6	6	6	6	5	5	5	6	5	3	5	4	4	4	5	5	(6)	4	5.1	1
15 / 1	6	7	5	6	6	6	6	4	(1)*	6	6	5	6	4	4	6	6	5	5	6	4	5	3	2	6	2	2	5	2	2	3	5.0	2
16 / 33	7	6	6	6	6	6	6	5	6	6	5	5	6	6	6	6	5	5	6	6	6	4	3	2	4	4	4	(6)*	5	(6)*	3	4.9	
17 / 18	6	6	5	6	5	5	5	6	4	6	5	5	6	6	6	5	5	5	5	6	5	4	4	5	3	4	4	4	5	5	3	4.9	
18 / 9	7	5	6	6	6	6	5	5	6	6	5	6	6	6	6	5	6	5	5	5	5	5	4	4	3	4	4	4	4	5	2	4.9	
19 / 21	5	5	6	5	6	4	6	6	5	6	5	5	4	6	6	5	5	5	5	6	5	5	5	5	6	4	4	4	4	5	4	4.9	
20 / 6	6	6	6	6	5	5	6	4	6	5	5	6	7	4	5	6	5	5	4	5	5	4	3	4	4	3	3	3	5	4	(6)*	4.8	
21 / 8	6	6	6	6	6	5	6	5	5	6	5	5	7	5	5	6	5	5	5	5	4	4	3	5	3	5	4	2	2	4	4	4.7	
22 / 29	6	6	6	7	5	4	5	5	6	5	5	5	6	5	5	4	5	5	5	5	3	(1)*	3	2	3	3	3	3	5	5	4	4.7	3
23 / 30	6	5	5	5	5	5	5	5	4	5	4	5	6	6	5	5	5	4	5	5	4	4	5	3	3	3	3	3	2	3	3	4.7	3
24 / 20	7	6	6	5	6	6	6	6	6	4	5	5	6	4	6	6	5	6	5	5	4	(1)*	2	3	3	2	3	4	4	3	3	4.7	
25 / 15	6	6	6	5	5	5	5	4	6	4	5	5	6	4	4	5	5	4	5	3	3	4	3	3	3	3	3	2	(1)	(1)	3	4.6	1
26 / 28	6	6	5	6	5	5	5	5	6	4	5	4	5	6	4	5	5	5	5	5	4	4	5	3	3	5	3	5	(1)	(1)	4	4.6	2
27 / 23	6	7	5	5	5	5	5	6	6	4	5	5	4	4	4	5	5	4	4	4	4	3	3	3	5	3	3	4	5	(1)	3	4.5	
28 / 4	6	5	6	6	5	5	6	4	5	4	4	5	4	5	4	4	4	4	5	4	4	(1)*	4	3	3	3	4	2	5	(1)	3	4.5	1
29 / 32	6	5	5	5	5	5	5	5	6	4	5	5	5	5	4	4	5	4	4	5	4	3	3	3	4	4	3	2	(1)	(1)	2	4.4	2
30 / 26	5	5	6	5	5	5	5	5	4	4	5	5	5	4	4	4	4	4	5	4	4	4	4	3	4	3	4	2	2	2	2	4.1	
31 / 22	5	4	5	5	5	5	4	4	4	4	4	4	4	4	4	4	4	4	4	4	4	4	4	3	3	3	3	2	3	1	2	4.0	1
32 / 27	6	5	3	4	5	5	3	5	6	5	3	3	5	4	3	3	3	3	4	4	3	1	3	3	2	2	3	3	1	1	2	3.8	3
33 / 31	4	5	3	4	5	3	3	4	2	5	3	3	5	4	3	4	4	3	4	4	3	1	(5)	3	(5)	3	3	1	1	(4)	(4)	3.3	3
Column Deviations:	1.2	1.1	1.0	.7	.6	.6	.6	.5	.5	.5	.5	.5	.5	.4	.3	.2	.2	.1	.1	0	-.1	-.5	-.6	-.9	-.9	-.9	-1.0	-1.2	-1.2	-1.3	-1.4	4.9	
n_j	1								2	2			2									5	1		3			3	5	7	2		31

166

TABLE 8.10.
ANALYSIS OF VARIANCE FOR THE 33 × 31 DATA OF TABLE 8.8.

Source	Degrees of Freedom	Sum of Squares	Mean Square	F
Column: dye base	32	337	10.5	12.8
Row: naphthol	30	571	19.0	23.2
$R × C$: residual	960	788	0.82	
Total	1022	1696		

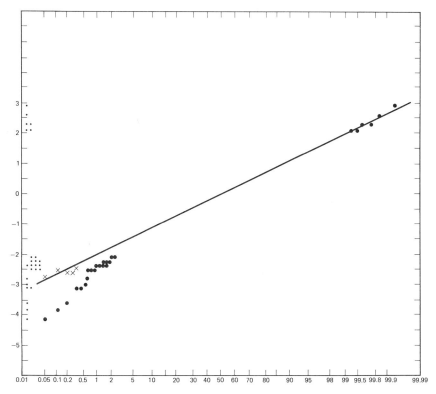

Figure 8.3 E.c.d. of 31 extreme residuals from Pfister's 33 × 31 · = data; x = after deletion of six points. See Table 8.9.

only to rule out six points as excessive—or, rather, deficient. There are reasons for believing that more than this number of cells in fact contain interactions.

As we look at Table 8.9 from a little distance, focusing now on the circled cells, we see that every circle except that at (1, 1) appears in a row or column with at least one other. We ask, then, how probable the observed row and column associations are, if the 31 largest residuals were simply dropped at random into the data table. The column n_i at the extreme right and the values n_j at the very bottom of Table 8.9 give us the required associations. Table 8.11 accumulates these as the numbers of rows N_i and columns N_j which contain one, two, etc., large residuals. Expected values of N_i and N_j have been computed from the binomial expansion of $(p + q)^n$, where $n = 33$ for rows and 31 for columns, $p = (1023 - 31)/1023 = 32/33 = 0.96$, and $q/p = 1/32$ for rows *and* columns. (The computation from the hypergeometric distribution gives so closely the same results that it is omitted here.) After the first term in the expansion, $p^n = A_0$, the succeeding terms were calculated by

$$A_i = A_{i-1} \frac{n - i + 1}{i} \cdot \frac{q}{p}, \qquad i = 1, 2, \ldots, 7.$$

TABLE 8.11.
NUMBER OF ROWS N_i AND COLUMNS N_j CONTAINING
$i, j = 0, 1, 2, \ldots, 7$ RESIDUALS WITH EXPECTED VALUES AND
CHI-SQUARE COMPONENTS

i, j	N_i	$E\{N_i\}$	χ_i^2	N_j	$E\{N_j\}$	χ_j^2
0	11	11.94	0.074	21	11.22	8.52
1	12	11.57	0.016	2	11.59	7.94
2	5	5.42	0.033	3	5.79	1.34
3	3⎫			2⎫		
4	0⎪			0⎪		
5	0⎬3	2.07	0.418	2⎬5	2.39	2.85
6	0⎪			0⎪		
7	0⎭			1⎭		
Sums	31	31.00	0.541	31	31.00	20.65

The results are almost too clear. For *rows* (i.e., for dye bases), expected values under the assumption of independence match the observations very well ($P = .90$). For *columns* (i.e., for naphthols), the two with *five* (columns 22 and 28 in Table 8.9) and the one with *seven* (column 30) take up far too many of the interactions, leaving too many empty columns and too few with only one large residual. The significance probability is less than .0005.

If we had the strength, we would replace all 31 large residuals by their values estimated from the unchanged row and column means, and redo the analysis of variance and the normal plot. The residual SS would be reduced by about 205 to 583 with (960 − 31) or 929 d.f. The revised MS residual would then be 0.628, and the "error std. dev." 0.79. But we do not have the strength.

As further subjective evidence that the circled values in Table 8.9 are interactive, we note that all columns with more than two excesses have the letter G in their specification, *and* that the three columns (22, 29, and 30) with five or more excesses all have LG in their names. These naphthols are chemically related, and this, for me, settles the matter. I have no doubt at all that these coupling agents did not operate additively on lightfastness with the dye bases indicated. From the point of view of the manufacturer of the naphthols, it must be advantageous to do the following:

1. Check the data for naphthols LG and L3G with dye bases 29, 30, 26, and 28 (in rows 22, 23, 26, and 27) since these are adverse interactions. The six combinations with asterisks are maximally disappointing.
2. Tell users that G, LG, and L3G give unusually favorable results (positive interactions) with the obvious dye bases.

I add only a sharp reproof to the statistician who would report the MS $(R \times C)$ as "random error." It is not. A less sharp rebuke, but still an admonition, should be given to those who simply report that this mean square is a mixture of random error and interaction. It is, but one can say exactly how, and so one should. We can even answer the question "Is the AB interaction due to A or to B?" It is widely stated that the 2fi's are symmetrical and that it is logically impossible to make a distinction between the statements "A operates differently at the different levels of B" and "B operates differently at the different levels of A." The interaction in our last case is due almost entirely to the naphthols and not to the dye bases. We all understand that the naphthols only operate nonadditively with some dye bases, but it is almost entirely the three naphthols L4G, LG, and LBG that produce the interaction. There is no trouble-making dye base. The observed interaction is not symmetrical in A and B.

Exactly analogous comments apply to most other cases when the nonadditivity is in more than one cell. In Table 8.2 all the interaction was in row 2. In Table 8.4 it was in variety 1 and in block I. In Table 8.5 (Immer on barley) the interaction was entirely due to *one* variety. I have not seen a case of *lack* of row or column association for 2fi's in data tables larger than 4 × 4.

These findings criticize implicitly my own overhasty drawing of the e.c.d. of the original residuals for this large table. I found only 6 excessive residuals from that plot, whereas I have found 31 by taking advantage of the *structure* of the table, that is, by the identification of the residuals, which is lost in the e.c.d. This is the same criticism that we made earlier of the use of half-normal plots in 2^5 experiments. Premature aggregation of visibly heterogeneous residuals is the hurried statistician's besetting sin, corresponding to the criticism made in Chapter 3 of the experimenter who generalizes prematurely.

8.7. GENERALITIES ON THE ANALYSIS OF BALANCED DATA

It will not have escaped the reader that little has been said in this chapter about the "statistical design of experiments." I take it for granted that any large balanced set of data, collected by a careful experimenter, is worth some study in its own right. The thorough statistician may retort that there is little value in this effort (especially for the data of Section 8.6) since no secure inferences can be drawn from such data. He will add that all the so-called tests of significance made are meaningless, and that all estimates are of unknown bias since the data were surely taken under nonrandomized conditions, probably in groups by columns or rows, by different, unspecified dye chemists, and perhaps on single batches of dye bases or coupling agents. (This statistician uses long sentences, doesn't he?)

I rely, more heavily than the skeptical statistician, on the intelligence and integrity of the producers of the data. It is true that all the "interactions" that I have uncovered may be due to defective randomization, to careless experimentation or record keeping, or to unstated changes in technique or technician. But the relatively small number of large deviations, and the resulting relatively small s of 0.91 including all interactions, make me believe that large interactions have not been overlooked, and that the row and column effects are consistent enough to be largely correct. It seems safe to me to conclude that the effects of these naphthols and dye bases on lightfastness are nearly all additive.

If I had been an advisor in the planning of this study, I would have recommended that at least six rows (if, as I suspect, rows were swept through in sequence) be repeated at random, *and* that the single cells showing residuals of, say, 2.5 in the first full replicate be repeated, changing technicians, material batches, or whatever conditions the dye chemist in charge thought desirable.

If a repeated row average came out far from its mate, say discrepant by 0.6 or more [we expect an s (row average) of $0.91/31^{1/2} = 0.163$], I would require still more row replication. Moreover, I could justify this to the dye chemists by the experience before us, rather than by appeal to the theory of statistics. (A further recommendation, which I have called partial replication, is discussed in Section 8.8)

Further checks are conceivable, without further data. We could list all residuals to the nearest tenth unit, in a table like Table 8.8, and study their homogeneity, omitting of course those circled in Table 8.9. This is left "as an exercise for the student." The reader may ask, rhetorically, "When does one cease analyzing the data?" My answer must be, "Only when time or money or strength runs out."

It has been a primary purpose of this chapter to give examples for analysis guided by—stimulated by—the data as they stand. There may be some indications of a future systematic program in these pages, but for the moment the emphasis should be on step-by-step study, aided by the usual tools and by any others the reader may think helpful. Is a transformation of the data desirable? Should four columns be removed from Table 8.7, and the rest reanalyzed? Does an analysis of variance of the absolute values of the residuals reveal anything? I welcome further suggestions.

8.8.   PARTIAL REPLICATION OF TWO-WAY LAYOUTS

Balanced (or partially balanced) incomplete block designs, for comparing v varieties, in b blocks, with k trials per block, can be written as two-way layouts with blocks for columns and varieties for rows. We can get an idea of k, for any given $R \times C$ (i.e., $v \times b$), by considering that we will need at least $R + C - 1$ degrees of freedom for separating out row and column parameters, and perhaps as many (surely not less than half as many) degrees of freedom for "error," in which we include, at the moment, interaction. We need, then, at least $3(R + C)/2$ observations. When R is about the same as C, this comes to roughly $3R$, and hence designs with $k = 3$ seem minimal and those with $k = 4$ better.

Such partial replications may find use in two different ways. They may provide an economical fraction of the full $R \times C$ table in cases where only rough screening of row and column parameters is envisaged. In situations like the one in Section 8.6, where more thorough calibration is required since the data will be used over a long period, a small fraction of the full $R \times C$ set might be added to a full replicate to get a well-scattered sampling of replication error.

It would be natural in planning the naphthol study to look for a supplementary set of, say, 36 varieties in 36 blocks of three (or four). Clatworthy [1973, page 274] gives a partially balanced incomplete block (PBIB) plan, LS68, that would spot five cells in each row and column of a 36×36 two-way layout. If no design of the size needed were available, I would not hesitate to scatter a set of the size required (roughly $3R$ in number) over the whole grid. These observations would be used primarily to get an error estimate. The gain through including them in the row and column parameter estimation would be trifling.

For smaller two-way tables, with b and v of the order of 5–10, numerous balanced incomplete block (BIB) and PBIB designs with $k = 3, 4$, provide suitable fractions. Computer estimation of row and column parameters and of all residuals in the observed cells is always much easier than hand calculation, but for plans in this size range the familiar formulae for adjusted means are not difficult. There is a list of five BIB plans in Davies [1971] and a somewhat larger list of references in Cochran and Cox [1957].

8.9. SUMMARY

The operations described above and summarized below are not to be taken as inflexible rules but only as a reflection of my own moderate success in analysing my clients' data.

1. If an $R \times C$ array has been replicated, learn about the mode of replication and study the empirical distribution of random error. If error appears Gaussian, nearly Gaussian, or Gaussian except for a few points, make and record reasonable revisions and estimate σ from the revised set.
2. If row and column categories are not orderable a priori, compute row and column sums and re-order table by decreasing sums both ways. This will help to detect localizable interaction and to make Tukey's G-test.
3. Compute, tabulate, and study the $\hat{\gamma}_{ij}$ (residuals) from the combined data. Tabulate row differences—if rows are longer than columns—and see whether the disparities spot the same cells as the $\hat{\gamma}_{ij}$. They will sometimes find more.*
4. If major $\hat{\gamma}_{ij}$ are in a few rows or columns, and are only a small set compared to $R \times C$, they should be revised to get more interesting, stable, and informative $\hat{\alpha}_i$ and $\hat{\beta}_j$ from the remainder of the data. The large residuals should of course all be reported, preferably after reflation to estimate the actual deviations from additivity. An analysis of variance table is useful when it partitions the total sum of squares into localizable interaction and into parts allocable to *consistent* row and column differences. One that uncritically spreads disturbances among row, column, and interaction SS's is not useful.

    The most satisfactory partitions are those that put all of the significant interaction SS into a few cells (or even into a single cell) except for a remainder which nicely matches that expected from random error.

*A paper on this possibility will appear shortly.

5. The greater the care exerted by the experimenter to sample just the population desired, the more valid are his findings. When objective randomization is assured, we are assured of a fairer allocation of sums of squares to effects and to random error. But even unrandomized data are often worthy of study, even if only to justify criticism. The result of such study may only be the spotting of a few values that the experimenter will want to repeat, or it may only find deplorable trends inextricably aliased with desired effects. Such findings can be useful educationally.

6. A large proportion of large $R \times C$ sets of data contain disturbances that are only technical or clerical errors. A large proportion of the remainder show disturbances that are in a few columns or rows (usually not in both). These must be pointed out to the experimenter.

7. When it makes experimental sense to sweep through rows or columns in sequence, the experimenter should be asked to repeat a few rows, perhaps those showing the largest and the smallest averages plus a small random subset. Even when this cannot be done, carefully worked trials are worth study. The data may provide internal evidence of coherence and of scientific value.

CHAPTER 9

The Size of Industrial Experiments

9.1. INTRODUCTION

The prime difference between industrial and academic experimenters seems to me to be that the former start with a budget, a staff, and a laboratory (or pilot plant or full-scale plant), and wish to improve a working system or to improve their knowledge of a working system. The latter start with a *problem* and then try to find the budget, the staff, and the equipment to cope with the problem, or at least with some manageable aspect of it.

Industrial research workers can often tell in advance about how many "runs" they will be able to make. This number, N_T, will depend heavily on the order in which runs can be made. It will usually be minimal if full randomization is required. In nearly all experimental situations some factors are hard to vary, whereas others, if not easy, are at least amenable to deliberate variation. Most industrial experiments are, then, split plot in their design. The total number of runs is largely determined by the number of combinations of the hard-to-vary factors that can be afforded.

It is usually not advisable to propose that all of the work be done in "balanced" or statistically planned sets. It is better to agree that some fraction, f, of the budget be set aside for "statistical experiments" so that the experimenters, perhaps all novices in statistical design, can watch the effectiveness of such designs and can evaluate their disadvantages, without the strain of being fully committed to a new and unfamiliar technique. In favorable situations, a fair value for f is 0.5.

Perhaps the hardest decision of the classical experimenter, in considering statistical designs, is committing himself to do a considerable piece of work,

175

which will require a substantial portion of his budget of time and funds, without being able to draw any quantitative conclusions until the whole set is completed. On the other side of the scale is the fact that the planning of larger blocks of work has often been recalled later by experimenters as a most valuable discipline.

As first mentioned in Chapter 3, all experimental campaigns should start with the preparation of an *influence matrix*. Using a row for each factor that is thought to influence any outcome, and a column for each response, the experimenter should make up a table that indicates the current state of opinion and knowledge about the system under study. In each cell of this table the experimenter can enter a summary of what he knows or guesses. Thus a $+$, or $-$ in the (i, m) cell would signal opinion about the direction of the effect of x_i, in the range considered, on y_m. If no opinion can be given, a "DK" (for "don't know") or an "i." (for "ignorance") should be entered. If a regression coefficient and its standard error are available, they can be written in. If the general shape of the relation is known, it can be sketched easily in a 1-inch square.

The independent variables, the factors, should be named and symbolized, and the range judged sensible for experimentation should be recorded. It is always important to know which factors are easiest to change and which hardest, and this information too should be set down. The degree of nesting to be used is determined almost entirely by the relative difficulty of varying the factors. Thus a change of feedstock to a large piece of equipment may require a long time to "line out," but a small change in pressure may be quick to equilibrate. In such a case the experimenter will surely want to run through several pressure levels before changing feedstock.

Assume now that the information matrix has been completed. Initial decisions have been made on the reasonable ranges over which to vary each important factor. The number of hard-to-vary factors n_H has been given. The number of runs N_H, each at different choices of levels of the hard-to-vary factors, has been roughly fixed for the budget period, perhaps for the project, in question. We do not propose to change any factor over a wide range. We are not exploring the outermost limits of operability of the system. We want only to get a generally valid picture of the conditions to which the system is sensitive, and of those toward which it is robust.

The cautious experimenter will rarely commit more than half his N_H, or half his time to deadline, to a single "statistical experiment." In this, he is of course entirely right. It often takes a number of additional runs, and may even require several more fractions of a factorial, to obtain satisfactory clarity about the operation of a system. Perhaps half of N_H should be reserved for this purpose. Hence the number of runs in the first set is guessed to be of the order of $fN_T/2 = N_H/2 = N_1$, say.

Experimenters usually have (or can develop), a rather large list of factors whose effects they would like to know. It will be necessary to choose n_H, the number of hard-to-vary factors, from this list to be of the order of $N_1/4$ to $N_1/2$, but in any case less than N_1. If this is not possible, some sort of group screening is called for. (See Watson [1961] and Patel [1962, 1963].) I expect n_H to be between 2 and 10, although the total number of candidate factors may be as large as 40.

If n_H is less than $N_1/2$, that is to say, if the number of hard-to-change factors is less than half the projected size of the first balanced set of runs, a good case can be made for committing all the statistical effort to a single design that will estimate all main effects separated from all 2fi's. Smaller plans of this type (called main-effect-clear plans, plans of Resolution IV, or four-letter plans) are given in Chapter 12, and larger examples are shown in Chapter 13.

When a few extra runs, say 4–8, can be spared beyond the balanced set just mentioned, the experimenter will have a better chance of disentangling and identifying any suspiciously large interaction strings. This operation is described in Chapter 14.

Only rarely must a plan of experimentation be set up in advance that will guarantee estimation of each main effect and of each 2fi aliased only with higher-order interactions. This will require at least 16, 22, 29, 37, 46, and 56 runs for 5, 6, 7, 8, 9, and 10 two-level factors, respectively. Such "two-factor-interaction-clear" or Resolution V plans are needed when the responses being studied demand long-term storage or the testing of equipment through many cycles of operation. In such cases, the possibility of experimenting or testing in stages is severely restricted.

9.2. EFFICIENCY AND ITS DEFICIENCIES

The term *efficiency* as used in statistics has little to do with the engineering or even with the commonsense use of the word. As E. S. Pearson has pointed out, we statisticians are oftentimes trapped by the honorific overtones of the words we choose. If an experimenter can get good estimates of seven main effects and all their 2fi's in 29 runs, it is stultifying to tell him that he can obtain "100% efficiency" of estimation only by doing 64 runs. The larger plan will give him the 28 desired estimates, of course with greater precision, but a large proportion of them will be estimates of zero. Thus this 100% efficient plan may waste 35 runs. Webb's saturated plan [1965] in 29 runs (consisting of (1), all 21 two-letter treatments, and all 7-six-letter combinations) is "only" 68% efficient, and so each effect is measured as precisely as if replicated about 10 times. Most experimenters had not dreamed that such precision was attainable with *no* replication.

A better measure of the usefulness of an experimental plan would take account both of the degree of saturation of the plan with useful effects and of the effective replication of statistical efficiency. Perhaps the product of the statistical efficiency and of the ratio (number of estimates/number of runs) can be called the *efficacy* or the *economy* of the plan. It would measure the efficiency *per useful degree of freedom.* Thus for Webb's 29-run plan just mentioned the economy would be 0.68 × 1.0 or 0.68. The economy of the standard 64-run plan for the same purpose would be 1.0 × 29/64 or 0.45. The fact that all the runs of the former are included in the latter is advantageous when sets of runs can be made and studied in sequence.

9.3. POWER AND SENSITIVITY

No mention has been made so far of the statistical power of the tests made to judge the reality of effects, or of the expected values of confidence-interval widths for parameters of interest. These both depend on advance knowledge of σ^2, the true error variance, and this is not usually known in advance. I still remember the few cases in which research workers gave estimates of the error variance that were supported by later work. Most commonly I have been given gross underestimates, based on the known high precision of the measuring instruments used on product properties.

My attempts to get good estimates of error variance by fitting equations to past data taken on the same system have occasionally been successful, but more frequently have been failures, and in both directions. The commonest cause of my underestimates is, I believe, my failure to spot replication degeneracy, some form of plot splitting, in old data. The commonest cause of my overestimates has been my failure to detect a few very bad runs. In historical unbalanced data it is often not possible to spot these. (For some examples of moderate success, see Daniel and Wood [1971].) The aberrant runs are less likely to show as having large residuals in poorly balanced data.

In working with experimenters who have already done large balanced multifactor tests, I have noted a third factor that has invalidated earlier variance estimates. The presence of an "external examiner" sometimes increases the care with which data are collected. This, in turn, may decrease the random error in new sets of data. A sort of reverse Heisenberg effect appears to operate. At the same time, undetected bad values in earlier tests must have invariably increased the estimates of error from these tests.

Finally, it has been my experience that research administrators have sometimes decided to proceed with an experimental campaign even after a power calculation has shown that a sequence of feasible size has low probability of detecting effects of the magnitudes desired. In such cases the objectives of the project were made more modest so that some experimentation could go forward.

The general conclusion seems to be that we must learn to do the best with what we have, using variance and confidence-interval-length estimates only as uncertain warnings of what may well be missed.

A more detailed treatment of the number of runs required must wait our discussion of fractional replication and of other incomplete factorials (Chapters 12 and 13). When the experimenter can see clearly the number of parameters (usually main effects and 2fi's) that he will want to estimate, he should then count on doing at least 1.3 times, and perhaps twice, that number of runs.

In risky summary of this discussion, an N_1 of 8 is minimal in industrial research. An N_1 of 16 is much commoner and usually more productive *per run*. Even such a set, however, frequently requires some augmentation, often with 8 more runs. Initial sets of 32 are less common but are of high yield when feasible. Finally, N_T's of 64–100 are not rare.

CHAPTER 10

Blocking Factorial Experiments

10.1. INTRODUCTION

We have been discussing factorial experiments ever since Chapter 3, but we now need to go into more detail about their subdivision into useful parts. The reader will surely know that factors are not necessarily all at two or three levels, and that there may well be more than two factors at more than three levels (as in Chapter 8).

Factorial experiments have been done since time immemorial by research engineers and by many others, but not under the name that Fisher gave them, and without using randomization, the factorial representation, or the general ideas of main effects and interactions. The data have usually been summarized as multiple plots, with all factors continuously variable.

An unreplicated factorial plan with n_A levels of A, etc., requires $N_T = n_A n_B n_C \ldots$ results for each response, and this number may be too large, or perhaps only so large as to raise concern about the stability of the system over the long time or wide area required. It is natural to look for sensible subgroups of the full factorial that will permit closer comparisons because of greater homogeneity within subgroups.

Such subgroups are called *blocks*. The classic example in agricultural experimentation is the subdivision of a single field in which a crop is grown into parts thought to be more homogeneous than the whole field. In industrial research, batches of raw material are frequently used to determine the size of blocks. In this way differences among batches are prevented from entering the error term, and hence greater precision is attained for the important comparisons, which are arranged to be estimable *within* blocks.

The practice of blocking for increased homogeneity is of course not confined to factorial plans. Any set of different treatments, varieties, or experimental conditions may be combined and applied to one batch of raw material, to one laboratory setup, or to one day's operation. If there is random allocation of treatments to experimental units, one *randomized block* has been accomplished. Confirmation by repetition in other randomized blocks provides maximum security in making inferences about systematic differences among the responses to different treatments.

Since it is often not possible to insert all treatments into each block, attention must be given to the selection of sub-subsets that will yield maximum precision in comparing the effects of treatments. The best subdivision will be that in which each treatment appears to be matched with every other treatment the same number of times. Such subdivisions are called *balanced incomplete blocks* (BIB for short). The BIB are best in the sense of giving maximum precision per observation, that is, highest efficiency. But although great ingenuity and effort have been expended in producing these designs, they often require an unacceptably large number of observations to attain full balance.

We are then forced to make a further compromise, by reneging on the requirement that all pairs be equally represented within blocks. These compromise plans are called *partially balanced incomplete blocks* (PBIB); and, if each pair of treatments appears *either* λ_1 *or* λ_2 times, together, they are called PBIB *with two associate classes*. The book by W. H. Clatworthy [1973] is certain to become the standard atlas of PBIB's.

An excellent discussion of the subdivision of factorial plans into randomized BIB's is to be found in Chapter 7 of Davies [1971] and is continued in Chapter 9. Part of the present chapter is a repetition, with admiration, of that discussion. However, a new system of blocking the 2^3 in pairs is given in Section 10.3.2 (none is given in Davies for this case). Such systems will be found useful when the effects of interest are of the order of $2\sigma_0$, where σ_0 is the error standard deviation *within* blocks. A new blocking system for the 2^4 in blocks of four, given in Section 10.4.2, makes it possible to acquire estimates of all but two 2fi's with full efficiency (the two exceptions with efficiency $\frac{1}{2}$). This is done by using different contrasts in blocking from the usual ones. Incidentally, or perhaps fundamentally, the standard dictum (that if P and Q

are confounded among four blocks, then PQ is also) is refuted by these designs.

10.2. THE SIMPLEST BLOCKING: THE 2^2 IN BLOCKS OF TWO

It might happen that natural blocks admit only two treatments (for example, twin goats or very small batches of some expensive raw material) and that only two two-level factors are under study. I do not consider this a likely case, but the fundamental principles of blocking are already involved and just escape degeneracy.

As in Chapter 3, we call the factors A and B and designate the four possible treatment combinations or experimental conditions as (1), a, b, and ab. We take two observations (1) and a, as a first attempt, and we look at the expected value of the only possible comparison *and* at the expected value of the sum of the two observations. By Yates's algorithm or by direct inspection,

(9.1) $E\{(1)\} = -A - B + AB +$ average level of block I,

(9.2) $E\{a\} = A - B - AB +$ average level of block I,

Hence, $E\{a - (1)\} = 2A - 2AB,$

and $E\{a + (1)\} = -2B + 2(\text{average level of block I}).$

So we see that the block difference estimates $2(A - AB)$ and the sum $2(\text{average level of block I} - B)$. In the jargon of blocking, AB is *aliased* with A, and B is *confounded* with the mean. The two new terms are, so far as I can see, synonymous.

There are only six possible pairings of the four treatment combinations, which are shown diagrammatically in Figure 10.1 and are designated here as

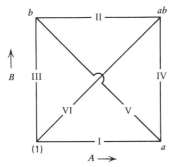

Figure 10.1   The six possible blocks of two for the 2^2.

TABLE 10.1a.

EXPECTED VALUES OF AND ESTIMATES FROM DIFFERENCES IN THE $2^2///2$

No.	Computation	$\frac{1}{2}$ Expected Value	Estimates from Block Differences and Sums
I_d	$a - (1)$	$A - AB$	$I_d + II_d = 2A_1$*
II_d	$ab - b$	$A + AB$	$II_d - I_d = 2AB_1$
III_d	$b - (1)$	$B - AB$	$III_d + IV_d = 2B_1$
IV_d	$ab - a$	$B + AB$	$IV_d - III_d = 2AB_2$†
V_d	$a - b$	$A - B$	$V_d + VI_d = 2A_2$
VI_d	$ab - (1)$	$A + B$	$VI_d - V_d = 2B_2$

* A_1 means "first estimate of A."

† AB_2 means "second estimate of AB."

I, II, . . . , VI. The expected values of the six block differences are shown in Table 10.1a and should be verified by the reader.

It comes as no surprise that at least three blocks of two must be done in order to get even the crudest estimates of A, B, and AB. Since the experimenter would only rarely do a 2^2 unless he was interested in the 2fi, AB, the two blocks V and VI are the least useful. Their internal comparisons do not contain AB. If the first four blocks are done, we see (and can verify computationally) that AB is estimated with half the variance of A and B. If all six blocks are done, then all three parameters are estimated equally precisely.

It should be apparent without computation that each of the three effects is estimated with efficiency $\frac{2}{3}$, since only four out of six block differences are used to estimate each one.

Since six differences are ultimately available, we must look for and make use of the three "degrees of freedom" not consumed in the estimation of the three parameters. These three are measures of (twice) the random error within blocks: they are found from $A_1 - A_2$, $B_1 - B_2$, and $AB_1 - AB_2$. This "intrablock variance" will be symbolized by σ_0^2.

The expectations of two times the six block means are given in Table 10.1b.

The new symbols, l_i, $i = 1, 2, \ldots, 6$, are parametric deviations of block means from the grand average, M. If the blocks can be viewed as a random sample of a population of blocks, the l_i may be used to estimate the variance of this population, which we will call σ_1^2. This may be useful for judging the effectiveness of the blocking. If σ_1^2/σ_0^2 is large, blocking has been in some degree effective. This ratio should be greater than 3 for real gains in precision.

Estimates of the l_i may be computed by correcting each block sum for the indicated effect, to get six estimates of $(M + l_i)$. The deviations of these from their average (which estimates M) are the values desired.

TABLE 10.1*b*.

EXPECTED VALUES AND ESTIMATES FROM BLOCK SUMS IN THE $2^2///2$

No.	Computation	$\frac{1}{2}$ Expected Value	Estimates from Block Sums
I_s	$(1) + a$	$M^* - B + l_1{}^\dagger$	$I_s + II_s = 2M + l_1 + l_2$
II_s	$b + ab$	$M + B + l_2$	$II_s - I_s = 2B - l_1 + l_2$
III_s	$(1) + b$	$M - A + l_3$	$III_s + IV_s = 2M + l_3 + l_4$
IV_s	$a + ab$	$M + A + l_4$	$IV_s - III_s = 2A - l_3 + l_4$
V_s	$a + b$	$M - AB + l_5$	$V_s + VI_s = 2M + l_5 + l_6$
VI_s	$(1) + ab$	$M + AB + l_6$	$VI_s - V_s = 2AB - l_5 + l_6$

* M stands for the parametric value of the mean of all six blocks.

† See text for definition of l_i.

The block parameters have obviously been assumed to enter additively into their expectations in Table 10.1*b*, and so any block-treatment interactions have been taken to be zero. If this assumption is in error, the three effect differences, $(A_1 - A_2)$, etc., will reflect these interactions. If all such differences are small (compared to A, B, AB), we need not worry about them at this stage. The practical reader will recognize that this pathetically small example, with only three d.f. for the estimation of σ_0^2 (and five for σ_1^2), is not recommended for actual use except in dire circumstances. It may be said, however, to be better than nothing. It is certainly better than the absolutely minimal design of blocks I, II, and III for estimation of A, B, and AB. But modesty will not guarantee usefulness. It is only the principles and computations that are to be remembered from this section.

Looking back at Tables 10.1*a* and *b*, we ask, "What is confounded with the within-block differences?" For blocks I and II we see that B is confounded with the difference between results in each, and that A and AB are estimable with variance $\sigma_0^2/2$. We have, then, deliberately lost B for the sake of better estimation of A and AB. *Ceteris paribus* in blocks III and IV, we lose A but get estimates of B and AB.

10.3. THE 2^3 IN BLOCKS OF FOUR AND TWO

10.3.1. The 2^3 in Two Blocks of Four

Here we come, for the first time, to a tolerable experimental situation in that we can "lose," that is, confound, a parameter that we do not usually cherish. The 3fi ABC is estimated by a contrast among the eight responses of the 2^3. If we put the four trials that have $+ABC$ in one block and the remaining four in the other, we have confounded ABC with the block difference. But we can estimate the remaining six parameters (the three main effects and the three 2fi's) with full efficiency and minimum variance. Thus we have

Block I	Block II
(1)	a
ab	b
ac	c
bc	abc

It is easiest to remember this partition by noting that the "even" trials are in one block. The symmetry is pleasant to see diagrammed (Figure 10.2). Here the circled vertexes specify block I. These appear nicely spaced in a tetrahedron, two at the high and two at the low level of each factor. We can discover

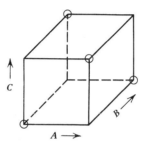

Figure 10.2   The two blocks of four for the 2^3, confounding ABC.

(again) just what each of the three obvious contrasts (e.g., the difference between the two results at high A and the two at low A) is measuring by the standard procedure of Yates on the 2^3, entering ± 1 for each of the four treatment combinations as it will enter the desired contrast. Thus for the A-contrast we have

Spec.	(0)	(1)	(2)	(3)	
(1)	-1	-1	0	0	
a		1	0	4	(A)
b		1	2	0	
ab	$+1$	-1	2	0	
c		1	2	0	
ac	$+1$	1	-2	0	
bc	-1	1	0	-4	(BC)
abc		1	0	0.	

We see that the A-contrast, called (A), has expected value $4(A - BC)$, measured with the within-block precision. The other two obvious contrasts measure $4(B - AC)$ and $4(C - AB)$. Since the other block yields estimates of $4(A + BC)$, etc., we can estimate all six parameters with full precision, that is, with variance $\sigma_0^2/8$. All six effects can be computed in one "Yates calculation," ignoring the result for ABC since it measures that interaction plus the block difference.

10.3.2.   The 2^3 in Blocks of Two

Blocks of two factorial treatments are often needed but are not frequently discussed in the statistical literature. They should always be considered when the variance between identifiable pairs is known to be a small fraction, say one third or less, of the variance of unblocked observations. There is usually some loss in variance efficiency in using blocks of two; but, as we will see, this efficiency factor is $\frac{1}{3}$ in the worst case (6 blocks of two for estimating six parameters in the 2^3) and is $\frac{2}{3}$ when $8-12$ blocks can be managed for the same factorial.

The familiar textbook example used to exemplify "partial confounding" will be skipped over lightly here. If the 2^3 is covered in the four blocks:

I	II	III	IV
(1)	a	b	ab
abc	bc	ac	c

the three main effects can be estimated with full efficiency, since each block gives an estimate of $A \pm B \pm C$, and all four differences can be combined to yield main effect estimates with variance $\sigma_0^2/8$. This seems to me shortsighted since all 2fi's have been confounded with block differences. Surely, if a factorial plan is contemplated, the 2fi's are of interest. It would be satisfying if the three 2fi's could all be estimated from a new set of four blocks, but this is not feasible by the usual method of confounding, which confounds some effect, say A, between the first pair of blocks and the second, and then confounds another effect, say B, between the pair I + III and the pair II + IV. If this is done, their product, AB, is inevitably confounded between the pairs I + IV and II + III, and there goes one of our desired 2fi's. Yates [1937] (and, following him, Cochran and Cox, Davies, and all other authors) recommends that another set of four blocks be done to get the missing 2fi's and that then still another set of four be carried out to attain balance, that is, equal variance for all estimates.

We tackle first the specification of *six* blocks of two that will permit estimation of the six first- and second-order effects in the 2^3, assuming ABC

to be negligible. Since we know that this cannot be done by using one set of confounded effects throughout, we employ a different set for each *pair* of blocks. We will want to estimate A and $(AB + AC)$ from blocks I and II, B and $(AB + BC)$ from blocks III and IV, and C and $(AC + BC)$ from blocks V and VI. For block I, since A and $(AB + AC)$ are to be estimated, B, C, and BC must be confounded with its mean, so the block must contain only (1) and a. Block II must estimate the same two parameters with one reversed sign and so contains bc and abc. Similarly blocks III and IV are to estimate B and the sum of its 2fi's and so must be aliasing A, C, and AC with their means. We work this out a little more formally below for blocks I and II, and summarize the whole plan and its efficiencies in Table 10.2.

TABLE 10.2.
THE 2^3 IN SIX BLOCKS OF TWO

I	II	III	IV	V	VI
(1)	bc	(1)	ac	(1)	ab
a	abc	b	abc	c	abc

Estimable
effects: A B C
$(AB + AC)$ $(AB + BC)$ $(AC + BC)$
$= z_1/4$ $= z_2/4$ $= z_3/4$

Effects:	A	B	C	AB	AC	BC
Efficiency factor:	$\frac{1}{3}$	$\frac{1}{3}$	$\frac{1}{3}$	$\frac{1}{2}$	$\frac{1}{2}$	$\frac{1}{2}$

Starting with the block differences for I and II; we have

$$E\{a - (1)\} = 2(A - AB - AC),$$
$$E\{abc - bc\} = 2(A + AB + AC).$$

Adding and then subtracting these two equations gives

$$E\{(abc - bc) + (a - (1))\} = 4A,$$
$$E\{(abc - bc) - (a - (1))\} = 4(AB + AC) = z_1.$$

We proceed similarly for the other two pairs of blocks, getting estimates of a main effect and of the sum of its two 2fi's and calling the latter z_2 and z_3, respectively. It is then easy to separate the three 2fi's:

$$8\widehat{AB} = z_1 + z_2 - z_3,$$
$$8\widehat{AC} = z_1 - z_2 + z_3,$$
$$8\widehat{BC} = -z_1 + z_2 + z_3.$$

The variance of each of these estimates is $3\sigma^2/16$. The minimum possible variance obtainable for a 2fi would come from six blocks around the edges of a 2^2 (*sic*!) and is $3\sigma^2/32$, so all of our efficiency factors for the three 2fi's are $\frac{1}{2}$. These factors are hardly to be boasted about, but they do provide estimates of the six effects and so have full "degree of freedom" efficiency; the plan is saturated with within-block estimates. The efficiencies are improved if eight or more blocks are manageable.

10.3.3.   The 2^3 in Eight and in Twelve Blocks of Two

Table 10.3 shows, in its first two lines, the generation and specification of a superblock containing four blocks of two. It provides four useful estimates, the maximal number for four blocks of two. The next two lines specify another set of four blocks of two, and now all six desired parameters are estimable with the efficiency factors given below in the table. When a third set of four can be added, we attain equality of efficiency for all six parameters, although the average factor has not changed.

TABLE 10.3.
THE 2^3 IN EIGHT AND IN TWELVE BLOCKS OF TWO

Alias Subgroups (Superblocks)	Estimable Parameters	Treatment Combinations			
$\pm A \pm BC \pm ABC$	B, C, AB, AC	(1)	a	b	ab
		bc	abc	c	ac
$\pm B \pm AC \pm ABC$	A, C, AB, BC	(1)	a	b	ab
		ac	c	abc	bc
$\pm C \pm AB \pm ABC$	A, B, AC, BC	(1)	a	c	ac
		ab	b	abc	bc

Effect:	A	B	C	AB	AC	BC	Average Efficiency
Efficiency (8 blocks):	$\frac{1}{2}$	$\frac{1}{2}$	1	1	$\frac{1}{2}$	$\frac{1}{2}$	$\frac{2}{3}$
Efficiency (12 blocks):	$\frac{2}{3}$	$\frac{2}{3}$	$\frac{2}{3}$	$\frac{2}{3}$	$\frac{2}{3}$	$\frac{2}{3}$	$\frac{2}{3}$!

In each of the sets-of-four blocks there is a "principal block" containing treatment (1) and one other. The other blocks can always be generated from the principal block by multiplication by any admissible treatment not already present. The principal block is always determined by the alias subgroup. The rule is as follows: Each member of the principal block has an *even* number of letters in common with each member of the alias subgroup (0 is taken as an even number). To be strict, all effects with an odd number of letters are aliased negatively with the mean of the principal block, and all with an even number of letters are aliased positively with the mean of that block.

The gains in estimability of these tiny designs are obtained by two changes from the standard plans. First, we have placed ABC in the alias subgroup, not among the estimable effects, since it is assumed to be 0 or at least negligible. Second, instead of using blocks that estimate only main effects, we have chosen combinations that give 2fi estimates as well. There seems to me to be little point in using complementary pairs [like (1) and abc] that can give no within-block information on 2fi's.

10.4. THE 2^4 IN BLOCKS OF TWO, FOUR, AND EIGHT

10.4.1. The 2^4 in Sixteen Blocks of Two

We have 10 parameters to estimate (four main effects and six 2fi's). Exhaustive trial dashes the hope that these might be arranged in two superblocks of eight blocks each. However, by using four superblocks of four blocks each, each superblock confounded differently, it is possible to reach an efficiency factor of $\frac{1}{2}$ for all main effects and for four 2fi's. The remaining two 2fi's are estimable with full efficiency.

Table 10.4 shows the four superblocks, the four parameters estimable from each, and the required treatments for each block. It is easy to see that the estimate of $(AC + BD)$ from I can be combined with the estimate of $(AC - BD)$ from II to separate the two components, and that similar operations can be carried out on all the other pairs.

TABLE 10.4.
SPECIFICATION OF SUPERBLOCKS AND BLOCKS OF TWO FOR THE 2^4

Superblock:	I	II	III	IV
Estimable effects:	A	A	B	C
	B	C	D	D
	$(AC + BD)$	$(AC - BD)$	$(AB + CD)$	$(AB - CD)$
	$(AD + BC)$	$(AD - BC)$	$(AD + BC)$	$(AD - BC)$
Specifications:	(1), ab	a, c	(1), bd	a, acd
	ac, bc	b, abc	ab, ad	b, bcd
	ad, bd	d, acd	$ac, abcd$	c, d
	$cd, abcd$	abd, bcd	bc, cd	abc, abd
Parity:	Even	Odd	Even	Odd
Generation:	$A, B \to ab$	$A, C \to a, c$	$B, D \to bd$	$C, D \to c, d$
		$= \{(1), ac\} \times a$		$= \{(1), cd\} \times c$

The reader is remined of the two-weight problem discussed in Chapter 1. It will be remembered that, when the sum and the difference of two weights (P and Q) can be observed, the variance of the estimates of the two weights is halved as compared to direct observation of each weight singly. Here too we estimate sums and differences, and so we name and then invoke the P-Q

PRINCIPLE. We set up one set of four blocks of two (superblock I) so that four estimates are obtained. It is not possible to arrange for $(A - B)$ and $(C - D)$ at the same time, so we settle for $A, B, (AC + BD)$, and $(AD + BC)$. Each of the four blocks of two must then have these four "parameters" in the expected value of its single contrast, but with different signs, so that the four desired combinations can be orthogonally estimated. Again, the P-Q principle is being applied, but now perhaps it should be called the P-Q-R-S principle, which is the optimal generalization for four weights.

Since we want $A, B, (AC + BD)$, and $(AD + BC)$ to be in the expected value of the block contrast, the parameters C, D, AB (and their products) must be confounded with the block mean. For the principal block, we must always take "odd" parameters with the minus sign, and so we generate the alias subgroup as

$$I - \underline{C} - \underline{D} + CD + \underline{AB} - ABC - ABD + ABCD,$$

where the three underlined terms are taken as generators.

Checking to make sure that the desired parameters are in the block difference, we multiply each term of the alias subgroup by A, say, and get

$$A - AC - AD + ACD + B - BC - BD + BCD$$
$$= A + B - (AC + BD) - (AD + BC) + ACD + BCD,$$

and so all is well.

To find the corresponding treatment combinations automatically, we can do reverse Yates (on a set of 0's and 1's), either on the members of the alias subgroup or on the string of effects just given. We find (1) and ab, and we note that each of these has an *even* number of letters in common with every member of the alias subgroup. It suffices to check only the three generators for evenness, as can easily be proved. Pains are taken to change the signs of *two* generators in each line, so that the superblock *in toto* will have only $ABCD$ confounded with its mean (plus the sum of all block differences). Here is the sequence of signs for each of the four blocks of two:

I	$\underline{C}$	$\underline{D}$	CD	$\underline{AB}$	ABC	ABD	$ABCD$	Treatment	Combinations
+	−	−	+	+	−	−	+	(1)	ab
+	+	−	−	−	−	+	+	ac	bc
+	−	+	−	−	+	−	+	ad	bd
+	+	+	+	+	+	+	+	cd	$abcd$

Superblock II is generated from a different set of four target parameters, now including $(AC - BD)$ and $(AD - BC)$ so that we can use the P-Q principle, and similarly for the two superblocks III and IV. A price is paid for this rather irregular blocking, however, and those seeing the penalty for the first time may find it too high. Only half the data are used (with full efficiency) in estimating 8 of the 10 desired parameters, while all are used to estimate AD and BC. Thus, in the old-fashioned sense, this plan has 50% efficiency for all but two parameters. We have accepted this deficiency in order to be able to estimate all 10 parameters in 16 blocks of two. These cannot be acquired by standard "balanced" methods without using twice as many blocks in all.

10.4.2.   The $2^4//4 \times 4$, That Is, the 2^4 in Four Blocks of Four

Standard dogma requires us to choose two factorial parameters, preferably higher-order interactions, *and their product*, which are assuredly negligible. We resist using $ABCD$ for one of the blocking parameters because its product with any other undesired parameter is either a main effect or a 2fi. If we choose any two 3fi's, their product is a 2fi, and so it might appear that one 2fi must be lost. This is the choice recommended by Cochran and Cox, by Davies, and by all other authors known to me, following Yates [1937]. To regain some kind of balance, *six* repetitions of the 2^4 are recommended by these authors, each using a different confounding pattern, and each losing a different 2fi. This may be feasible in agricultural experiments when 96 plots are needed to attain adequate precision. But even two repetitions, losing one 2fi in each, may suffice. We would then have full efficiency on all but the two blocking 2fi's, which would be estimated with efficiency factor $\frac{1}{2}$.

Table 10.5 gives a blocking scheme for a single replicate of the 2^4 which permits estimation of all 10 main effects and 2fi's, 8 with full efficiency, but BD and CD with efficiency $\frac{1}{2}$. The four defining contrasts* are as follows:

> I. $I - ABC - ABD + CD$.
> II. $I - ABC + ABD - CD$.
> III. $I + ABC - ACD - BD$.
> IV. $I + ABC + ACD + BD$.

TABLE 10.5.
THE 2^4 IN FOUR BLOCKS OF
FOUR (ALL 2fi's ESTIMABLE)

I	II	III	IV
(1)	abd	a	abc
ab	ac	abcd	ad
acd	bc	bd	b
bcd	d	c	cd

---

* I have not used the name "defining contrasts" before. It designates the sum of the terms in the alias subgroup and hence the list of parameters that are confounded with the mean of one block. All other aliases can be found by multiplying this string by any effect not in the string.

The reader should be warned that, although in this case a single Yates's computation on the 16 results will give correctly the estimates of eight of the desired parameters, a different computation—using only the half of the data in which it is not confounded—will be required for *BD* and for *CD*.

I hear a voice muttering, "But this is merely partial confounding." I mutter back, "Yes, but not 'merely,' since the plans produced have long been needed but not offered."*

There are 3 d.f. within each block and hence 12 in the four blocks, so there must be 2 d.f. for error. These are *BCD* and *ABCD* and are estimated with full efficiency.

10.4.3.   The $2^4//8 \times 2$

It seems inevitable to choose *ABCD* for confounding. This divides the 2^4 into an "even" and an "odd" half. It is of some interest to note that the four main effects (each along with the complementary 3fi) can be estimated from either block.

10.5.   THE BLOCKED 2^5

For two blocks of 16, the 5fi *ABCDE* is clearly the parameter to lose. For four blocks of eight, any two 3fi's with one letter in common should be chosen, since their product will then be a 4fi. If the experimenter suspects that certain 2fi's may be large, their letters should be split between the two 3fi's.

There must be subdivisions of the 2^5 into blocks of four, and even of two, but they are not given here. Enough has been said about the 2^3 and the 2^4 to show what rules to follow and which to circumvent.

10.6.   THE 3^2 IN BLOCKS OF TWO AND THREE

The authors of Davies [1971] have in their Chapter 9, Sections 5–9, and in Appendix 9F and G, given an excellent discussion of the natural ways to block the 3^2, the 3^3, and the 3^4 in blocks of 3, 9, and 27. Since these arrangements have been exactly duplicated in dozens of textbooks, I decline to repeat them here, but only recommend their use when blocks of three are the natural ones to use.

The partitioning of the interactions originated by Yates is described in all texts as uninterpretable. A little space is taken here to rectify this judgment.

The *I* interaction pair of degrees of freedom is defined [Yates 1937, page 95; Cochran and Cox 1957, page 193; Quenouille 1953, page 120; Davies 1971, page 399] by the three "diagonal sums" of the observations in a 3^2.

---

* See Youden [1961] for a different constructive use of partial confounding.

Labeling the nine cells as follows:

	b_0	b_1	b_2
a_0	1	4	7
a_1	2	5	8
a_2	3	6	9

we have the definitions

$$I_1 = 1 + 5 + 9, \qquad J_1 = 1 + 6 + 8,$$
$$I_2 = 2 + 6 + 7, \qquad J_2 = 2 + 4 + 9,$$
$$I_3 = 3 + 4 + 8; \qquad J_3 = 3 + 5 + 7.$$

Yates partitions the interaction sum of squares into two pairs of two d.f., one among the I, the other among the J, but does not give orthogonal contrasts corresponding to individual degrees of freedom.

Taking the (only) two natural contrasts, I define

$$U_1 = 2I_1 - I_2 - I_3 \qquad \text{and} \qquad V_1 = 2J_3 - J_1 - J_2;$$
$$U_2 = I_2 - I_3 \qquad \text{and} \qquad V_2 = J_1 - J_2.$$

The U_i and V_i $(i = 1, 2)$ are free of main effects and of the grand average and are therefore solely functions of the 2fi, that is, the $\hat{\gamma}_{ij}$ of the factorial representation. Using now the $\hat{\gamma}_{ij}$ or residuals in the 3^2 instead of the observations, we can list the *multipliers* of each residual in a standard 3×3 array to represent each U and V.

$$U_1 = \begin{vmatrix} 2 & -1 & -1 \\ -1 & 2 & -1 \\ -1 & -1 & 2 \end{vmatrix} \cong \begin{vmatrix} 3 & & \\ & 3 & \\ & & 3 \end{vmatrix}, \qquad U_2 = \begin{vmatrix} 0 & -1 & 1 \\ 1 & 0 & -1 \\ -1 & 1 & 0 \end{vmatrix};$$

$$V_1 = \begin{vmatrix} -1 & -1 & 2 \\ -1 & 2 & -1 \\ 2 & -1 & -1 \end{vmatrix} \cong \begin{vmatrix} & & 3 \\ & 3 & \\ 3 & & \end{vmatrix}, \qquad V_2 = \begin{vmatrix} 1 & -1 & 0 \\ -1 & 0 & 1 \\ 0 & 1 & -1 \end{vmatrix}.$$

From these it follows that:

$$U_1 + V_1 = (A_Q B_Q), \qquad U_1 - V_1 = 3(A_L B_L);$$
$$U_2 - V_2 = (A_Q B_L), \qquad U_2 + V_2 = (A_L B_Q),$$

where the parentheses on the right mean "the standard integer contrast among the residuals."

We can now put into words what the I and J components mean. $U_1 (= 2I_1 - I_2 - I_3)$ is a single-degree-of-freedom contrast which will respond to a diagonal ridge (or trench) in the responses in the direction \. $V_1 (= 2J_3 - J_1 - J_2)$ will be large only when a diagonal ridge (or trench) goes in the opposite direction, /. U_2 and V_2 are simply orthogonal remainders, representing lack of fit to one or the other ridge. If both U_1 and V_1 came out large, we would of course prefer to represent their sum as $A_Q B_Q$.

I leave to fresher minds the working out of the corresponding extensions to the 3^3, and so forth. The actual partitions are given in all four references cited above.

We try as always to reach a reasonable compromise between the number of blocks required for exact balance (equality of variance of all estimates) and the number of parameters to be estimated. For the 3^2 there are of course eight parameters to be estimated from within-block comparisons. If we require all 36 blocks of two ($9 \times 8/2 = 36$), we will reach exact balance. Remembering that we want good row and column comparisons, we start with the 12 pairs that compare each cell with its edge neighbor. These will give good estimates of the comparison of row (or column) 1 with row (or column) 2 and of row (or column) 2 with row (or column) 3. Numbering the cells as follows:

$$
\begin{array}{ccc}
1 & 2 & 3 \\
4 & 5 & 6 \\
7 & 8 & 9
\end{array}
$$

we are proposing, then, these blocks: 12, 23, 36, 25, 14, 45, 56, 69, 58, 47, 78, 89. The 12 differences can be used to estimate the row and column differences and the interaction parameters. There will be 4 d.f. for within-block error. If this is not deemed sufficient, the natural augmentation is to the comparison of row (and column) 1 with row (and column) 3; hence we add blocks 17, 28, 39, 13, 46, 79. This will give six more d.f. for error and notably improved efficiency and precision in all estimates. The 18-block plan appears as Design LS1 on page 260 of Clatworthy [1973].

10.7. DISCUSSION AND SUMMARY

There has been insufficient warning in the preceding sections about the hazards of blocking. It has been assumed that the true effects and their interactions are the same in all blocks. In the jargon of this chapter, it has been assumed that blocks do not interact with factors.

In serious experimentation, however, this is almost never known beforehand. The relative success of blocking must depend on the fact that interactions are commonly rarer than effects, at least when factors are varied over the ranges actually chosen by experimenters. Extensive retrospective reviews (by Kempthorne [1952] and others) have verified that, in experimental agriculture, block-factor interactions must have been small. This assumption can be partly tested by using a larger number of blocks than the absolute minimum for estimability, so that several effect estimates can be made and then compared.

Three factors have motivated this rather lengthy discussion of blocking. The first is the need to counter the inveterate habit of many experimenters who believe that the only way to guarantee "controlled experiments" is to use a large part of their experimental time and effort in repeated measurement of "standards." Blocking provides, as Fisher never tired of emphasizing, local control, that is, within-block comparisons that largely eliminate the long-term drifts of many experimental systems. The second factor is my desire to shake up a little those statisticians who take it for granted that no improvements in blocking factorials are to be expected. The third is the need to prepare the reader for "fractional replicates," which are merely rather large blocks, used singly to estimate lower-order parameters aliased with negligible higher-order ones.

CHAPTER 11

Fractional Replication—Elementary

11.1. INTRODUCTION

A fractional replicate is simply a block of a full factorial plan. A chosen set of effects, usually interactions that can safely be judged negligible, is confounded with the mean. This set, together with the identity I, is the alias subgroup.

Every effect in a fractional replicate is biased by the complement of each member of the alias subgroup. It is important therefore that all these complements be interactions that are either safely negligible or easily identifiable.

There are two reasons for proposing a fractional replicate instead of a full replicate:

1. The effects of the factors of primary interest can be examined over a wider range of conditions than would otherwise be possible.
2. The number of runs required to investigate a given number of main effects and 2fi's is decreased.

The principal disadvantages are:

1. Too few degrees of freedom may remain for testing for the multifarious varieties of lack of fit.
2. The vulnerability of fractional replicates to the usual hazards of experimentation—wild values, interchanged observations, inoperable or unattainable test conditions—is greater than that of whole replicates.

After discussion of these gains and hazards, we will make suggestions for reaching an acceptable balance. These proposals will be largely on the side of conservatism, judging only relatively large effects to be real, and reserving about half of the available degrees of freedom for study of the data.

11.2. FRACTIONS OF 2^p

The 2^p plans have been praised so extensively in earlier chapters that the reader must be convinced of their superiority. Why, then, do we rock this steady boat by putting forth modifications, especially if they are riskier, require more restrictive assumptions, and are harder to analyze?

Every 2^p factorial experiment *is* a fraction of a larger 2^P $(P > p)$ in which some factors have not been varied, but have deliberately been held constant, possibly not at their best levels. Thus any 2^2 is a half of a 2^3, the third factor being any condition held constant during all four runs of the 2^2. There may be better halves. These will be considered below. Every 2^2 is a quarter replicate, a 2^{4-2}, too.

In other words, there are usually more than p factors in each experimental situation. Very naturally, the experimenter's censoring judgment has been exercised in choosing the factors that seem worth studying and feasible to vary. He needs to be informed that some of the factors he has chosen *not* to vary might well be varied, and their effects noted, with no great increase (sometimes with no increase at all) in the number of runs required.

Put still another way, we may be able to broaden the base of our inferences about the effects of the p important factors by varying some other factors which "probably" produce no effects. We do not know that these latter factors are uninfluential; we only hope that they are. If our data show that they are indeed negligible, a point has been gained. If, on the other hand, one or more of them do influence results, an even more important fact has been learned.

There are of course nice matters of judgment in any decision to vary more factors. Shall we try to find out more about the more restricted system, or to find out less about the wider set of conditions implied in varying more factors?

11.3. SOME OVERSIMPLE FRACTIONAL REPLICATES

11.3.1. One Run

The single run is made at the "low" levels of three factors. (The naming of factor levels is generally a matter of nomenclatural convenience.) We specify the conditions of this run by the symbol (1). To be neat, we should specify the outcome—the response of the system to the run made at condition

(1)—by some different symbol, say $y_{(1)}$. But it is usually typographically convenient, when no ambiguity is likely, to designate the value of the response also as (1).

The factorial representation of the result (1) shows only how its expected value is related to the expected value of the average response to the full factorial. Thus, if three factors are being held at their lower levels, we can write

(11.1) $$E\{(1)\} = M - A - B + AB - C + AC + BC - ABC,$$

just as the first line of Table 5.2 shows. The expected value of the result at (1) is quite obviously biased from M by any of the seven factorial effects which are not zero.

Although this "experiment" has been called oversimple, it represents the commonest of all tests or experiments. A run is made under *some* conditions, mainly to see whether an interesting or an acceptable "yield" is obtained. If we want some yield but get none, this is often a crucial finding. Even getting a very low yield is informative. When this happens, the experimenter may in his next run try conditions as different as possible from those tried at first.

11.3.2. Two Runs

Suppose the experimenter now tries *abc*. He has changed the levels of all three factors, guessing that all three may influence the response in the same direction. The last line of Table 5.2 shows how the factorial effects hit this run. The *difference* between the two responses has the expected value:

(11.2) $$E\{abc - (1)\} = 2(A + B + C + ABC).$$

Now only four effects are aliased. We have done one fourth of the 2^3 that would be required to give us estimates of all eight factorial effects unaliased with each other. We can call our two runs a quarter replicate of the 2^3 and symbolize this as a 2^{3-2}.

11.4. THE HALF REPLICATE, 2^{3-1}

We cannot conceivably get separate estimates of the main effects of factors A, B, and C unless we do four runs. Let us ask first a simple question, "Where should we place four runs in a 2^3?" Intuitively we say that they should span the cube as well as possible. This forces us to one of the two complementary blocks of four of Section 10.3.1 (diagrammed in Figure 10.2). As we saw in that section, ABC is confounded with the block difference. As we should see now, the mean of the *odd* block is aliased with $+ABC$, and that of the even block with $-ABC$. If this is not clear, the reader should *either* consult Table 5.2 (which will show that the odd treatment combinations are all on the plus side of ABC, all the even ones on the negative side), *or* carry through the forward Yates computation, using $+1$ for the four even combinations and -1 for the rest, *or* do forward Yates on a set of $+1$'s in the even positions, to find that the mean and $-ABC$ are measured by the average of the four runs.

Just as in Chapter 10, here we can multiply the defining contrast by any effect not present to display all factorial effects aliased with the multiplier. Thus for block I we have

$$E\{\bar{y}\} = I - ABC,$$
$$E\{(A)\} = 4(A - BC),$$
$$E\{(B)\} = 4(B - AC),$$
$$E\{(C)\} = 4(C - AB),$$

where $\bar{y}$ is the block mean, and (A) is the contrast $ab + ac - (1) - bc$, etc.

A more deliberate way to produce the 2^{3-1} may prove more illuminating. It is the approach used by the authors represented in Davies [1971] and is beautifully explained in Box and Hunter [1961]. Table 11.1 gives at the left the standard transformation matrix for the 2^2.

TABLE 11.1.
IDENTIFICATION OF $\pm AB$ WITH C TO PRODUCE
A 2^{3-1}

Spec.	T	A	B	AB	C_1+	C_2-
(1)	$+$	$-$	$-$	$+$	c	(1)
a	$+$	$+$	$-$	$-$	a	$a\underline{c}$
b	$+$	$-$	$+$	$-$	b	$b\underline{c}$
ab	$+$	$+$	$+$	$+$	$ab\underline{c}$	ab

We surely want to estimate A and B, but we ask whether AB is needed. If it is not, the AB-contrast can be used to measure the effect of C. We have two choices: we can assign the first and fourth runs to high C, as in the column headed C_1; or we can use the other two runs for high C, as in C_2.

We have, in effect, set $AB = C$ in C_1, and $= -C$ in C_2. More correctly, we use the AB-contrast to measure $(C + AB)$ in C_1 and $(C - AB)$ in C_2. I am happy to report that Box and Hunter express defining contrasts as the sums of aliased terms, as I have, instead of the confusing "equalities" of earlier works. (Davies [1971], Brownlee et al. [1948], and Finney [1945] use expressions like $A = -BC$ to indicate that $(A - BC)$ is measured by the A-contrast.)

11.5.   THE 2^{4-1}

The 2^{4-1} has 2^3 or eight runs. We therefore identify the $-ABC$-contrast in the standard 2^3 with the new factor, D. The specifications and aliases are as follows:

Specs.	Aliases
(1)	$I + ABCD$
$a\ \ d$	$A + BCD$
$b\ \ d$	$B + ACD$
ab	$AB + CD$
$c\ \ d$	$C + ABD$
ac	$AC + BD$
bc	$BC + AD$
$abc\ \ d$	$ABC + D$

This alias pattern has a qualitative advantage over those shown earlier in that main effects are now estimated with no bias from 2fi's. The latter are of course aliased in pairs.

We will call plans of this sort four-letter plans or plans of Resolution IV (following Box and Hunter). Plans which force main effects to be aliased with 2fi's (like the 2^{3-1} of Section 11.4) are called, by various writers, three-letter plans, plans of Resolution III, or main effect plans.

The data and column 3 of the standard computations for the example which follows are taken from Davies [1971, pages 454–457, 491]. "The error variance was known to be about 4.0." Using $\bar{y}, (A)$, and (B) in the "reverse Yates" shown in panel b of Table 11.2, we find the fitted values Y_1 and the residuals d_1.

The conclusions appear simple and straightforward. The residual MS_1 of 6.0 with five d.f. is compatible with the given error variance of 4.0.

A suspicious mind would notice that the two largest residuals (-3 and $+4$) appear at the same experimental condition (a, since C and D are without effects). Since the true standard deviation is stated to be 2.0, the standardized range of this pair is 7.0/2.0 or 3.5, and this has a P-value of 0.025. This might suffice to raise one eyebrow, but not two.

We note that the signs of the ordered residuals (except for the first, which is 0) are those of $-AC$. Since AC is 12/8 or 1.5, we can revise the Y-values shown by this amount—and hence the residuals d_1 in the opposite sense—to get new residuals d_2, also shown in the table. The new residual mean square, residual MS_2, is 12/4 or 3, so we do not seem to be overfitting.

Our conclusions about effects are the same as Davies'. The A and B-effects are undoubtedly real and closely additive. The AC interaction is "probably" present, probably positive, and probably less than 2×1.5 or 3.

TABLE 11.2.

Panel *a*. Forward Yates on Davies' 2^{4-1}

Specs.	y	(3)	Aliases	$Y_1(\bar{y}, \hat{A}, \hat{B})$	d_1	Including $\widehat{AC} = 1.5$	d_2
(1)	107	952*	T	107	0	—	+1.5
ad	114	42*	A	117	−3	+	−1.5
bd	122	56*	B	121	+1	—	−0.5
ab	130	−2	$AB + CD$	131	−1	+	+0.5
cd	106	6	C	107	−1	+	+0.5
ac	121	12	$AC + BD$	117	+4	—	+2.5
bc	120	−6	$BC + AD$	121	−1	+	+0.5
abcd	132	−4	$ABC + D$	131	+1	—	−0.5

Panel *b*. Reverse Yates on $T, (A), (B)$

Specs.	Effects	(1)	(2)	$(2) \div 8 = Y_1$	d_1
(AB)	0	56	1050	131	−1, −1
(B)	56	994	966	121	+1, −1
(A)	42	56	938	117	−3, +4
T	952	910	854	107	0, −1

From d_1, residual $MS_1 = 30/5 = 6.0$.
From d_2, residual $MS_2 = 12/4 = 3.0$.

* Judged significant

11.6.   A NOTE ON THE CHOICE OF $I \pm ABCD$

There is an apparent asymmetry between the two half replicates, $I + ABCD$ and $I - ABCD$! Suppose that all four main effects were present, were large, and were of about the same size. If *one* is of opposite sign to the other three, then, aside from random disturbances, we will see a pattern in the data taken according to $I + ABCD$ of the form shown in Table 11.3, panel *a*.

If the same experimental situation had been studied using the other half replicate, we would get results with a pattern like that in panel *b*. We now see six data values at the mean, one well above and one equally far below. This pattern is of course entirely acceptable mathematically and will give the right estimates of all four effects if all the data are all right. But the suspicious experimenter or data analyst, noting that all but two of his values are nearly the same, might be tempted to revise these two, or at least to doubt their validity. This would of course be disastrous, since then no effects at all

TABLE 11.3.

CONSTRUCTED DATA FROM THE TWO HALF REPLICATES

$I \pm ABCD$: $|A| = |B| = |C| = |D|$

Panel a. $I + ABCD$

$D + ABC$	1	1	2	2	$abcd$
BC		1	0	-2	bc
AC		-1	0	2	ac
C	1	1	-2	2	cd
AB		-1	0	-2	ab
B	-1	1	2	-2	bd
A	1	-1	2	2	ad
T		-1	0	-2	(1)

Panel b. $I - ABCD$

$-D + ABC$	-1	-1	0	0
BC		1	0	0
AC		-1	2	4
C	1	1	-2	0
AB		1	2	0
B	-1	1	2	-4
A	1	-1	0	0
T		-1	0	0

would be found. A revision of two values in the data of panel a will not remove all effects.

If the experimenter knows or is able to guess the signs of his effects, he can run through the little reverse Yates required, to see whether he is likely to get a pattern like that of panel a or that of panel b. He can then change the names of the levels of one factor if necessary and so give himself a better chance of seeing some effect in every data point, instead of just in two out of eight.

We close this chapter with a warning. We have so far found no fault with fractions-of-2^p plans, but this is a consequence of our not having put these small plans to sufficiently severe test. Such tests will be made in the following chapter.

The mature reader will understand that the small fractions, 2^{3-1} and 2^{4-1}, are too heavily saturated and too imprecise for most practical uses.

CHAPTER 12

Fractional Replication—Intermediate

If you want to make sense, you had better take all important factors into account.   PAUL GOODMAN.

12.1.   INTRODUCTION

There are several excellent atlases of 2^{p-q} plans for $4 < p < 16$ and for $16 < 2^{p-q} \leq 256$. They will all be praised in this chapter. The impatient experimenter or statistician may feel that he can bypass this plodding exposition and proceed to these references. So he can, but not safely. I have seen many—perhaps a hundred—ill-conceived, misleading, and therefore very expensive fractionally replicated experiments. The commonest defects are the following:

1.   Oversaturation, that is, too many effects demanded for the number of trials used.

2. Overconservativeness, that is, too many observations for the desired estimates.
3. Failure to study the data for bad values.
4. Failure to take account of all the aliasing. .
5. Imprecision, that is, insufficient hidden replication, due principally to ignorance of the error variance and of its constancy.

We will first look at fractions of some large factorials. Then, after this cautionary tale, some higher fractions, $2^{p-2,3,4}$, will be discussed. Finally the published listings for two- and three-level plans will be reviewed.

12.2. HALVING THE 2^5

12.2.1. Yates's 2^5 on Beans

We have already divided this 2^5 (in Section 7.2.3) into two half replicates, using the defining contrasts $I \pm DNPK$ as shown in Table 7.4. This was done, the reader will remember, to study the responses in blocks I and III, which

TABLE 12.1.
CONTRAST-SUMS FROM TWO 2^{5-1}'S FROM YATES'S 2^5 ON BEANS
(BLOCKING: SDP, SNK, $DNPK$)*

	$I - SDNPK$			$I + SDNPK$	
Specs.	Effects	Aliases	Specs.	Effects	Aliases
(1)			k		
sk	-87	S	s	-41	S
dk	117	D	d	75	D
sd	79	SD	sdk	1	SD
nk	55	N	n	-5	N
sn	41	SN	snk	13	SN
dn	69	DN	dnk	13	DN
$sdnk$	19	$SDN - PK$	sdn	11	$SDN + PK$
pk	-25	P	p	-59	P
sp	5	SP	spk	43	SP
dp	45	$DP - SNK^*$	dpk	-53	$DP + SNK^*$
$sdpk$	-129	$SDP^* - NK$	sdp	-59	$SDP^* + NK$
np	-29	NP	npk	-53	NP
$snpk$	41	$SNP - DK$	snp	-23	$SNP + DK$
$dnpk$	-59	$DNP - SK$	dnp	77	$DNP + SK$
$sdnp$	-65	$SDNP - K$	$sdnpk$	55	$SDNP + K$

had on average smaller residuals and yielded, he will not forget, a dusty answer. We will now divide Yates's 2^5 into the two halves that would be most natural a priori, ignoring for the moment the original blocking. We use then the defining contrasts $I \pm SDNPK$ and find effects as contrast-sums, listed in Table 12.1 and plotted in Figure 12.1.

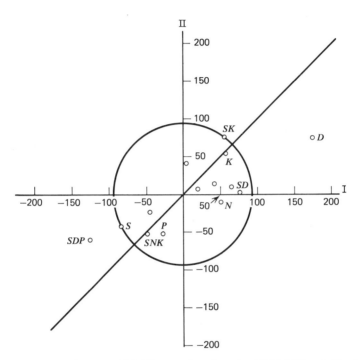

Figure 12.1 Contrasts from $I - SDNPK$ versus those from $I + SDNPK$ (Yates's 2^5 on beans).

I have used a t-value of 2.63—tabled at the 0.975 level and 10 d.f.—along with a pooled estimated standard error of contrast-sums of 36.2, to get a radius of 2.63×36.2 or 95.2 for the circle shown in the figure. Each half replicate gives roughly 10 d.f. for error, and even though the two halves do not give exactly the same s, I have taken the easy way out to avoid drawing ellipses. The circle in Figure 12.1 does appear to separate the largest effects (S, D, and SDP blocks) from the ruck, but the interesting finding discussed in Chapter 7—that $S \approx K \approx SK$—would barely be suspected in the half called I, and would be missed entirely in II.

We could have read in Yates's pamphlet [1937, page 30] that "The experiment is not of high precision, being of only 32 plots and having a high standard error per plot (beans have at Rothamsted proved a very variable crop) but in combination with other similar experiments it should provide useful information . . . ." We have then really gone in the wrong direction in trying to interpret half of this experiment and have done so only as an exercise, not as a suggested economy. The 2^{5-1} is by far the commonest fractional replicate actually done, so we should be warned that much may be missed.

12.2.2.   Davies' 2^5 on Penicillin

This experiment (see Section 7.3) was indeed blocked on $ABCDE$, and so each block was a 2^{5-1} of the sort that one might do singly.

The usual computations give the contrasts shown in Table 12.2 and the correlation plotted in Figure 12.2. Here the two halves find the same main effects significant. The circle drawn is of radius $2.63 \times 68 = 179$. The

TABLE 12.2.
TWO 2^{5-1} FROM DAVIES' 2^5

$I - ABCDE$		$I + ABCDE$	
-298	A	-264	A
-58	B	76	B
-104		-86	
228	C	-286	C
-134		-60	
26		116	
-56		14	
-26	D	58	D
-124		-36	
36		-68	
122	$ABD - CE$	-214	$ABD + CE$
10		-42	
-20		76	
-84	$-AE$	0	
398	$-E$	-270	E

troublesome *CE* interaction is large in one half, but much smaller in the other, and so might well have been missed. The value at *abcd*, called bad in Section 7.3.3, cannot be detected in its 2^{5-1}.

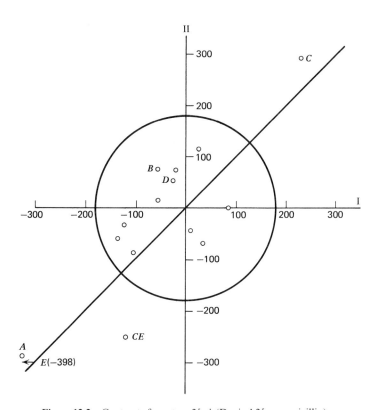

Figure 12.2 Contrasts from two 2^{5-1} (Davies' 2^5 on penicillin).

12.2.3. Rothamsted's 2^5 on Mangolds

As in Sections 12.2.1 and 12.2.2, we give the usual contrast-sums and their aliases for the two half replicates in Table 12.3.

As Figure 12.3 makes clear, the effects of *S*, *D*, and *N* are detected and well matched in both halves; the 2fi *KN* is a false positive in the principal block, but not in the other one. The 2fi *SK* was judged significant in the original 2^5 and is easily spotted in both halves.

TABLE 12.3.
Two 2^{5-1} from Rothamsted 2^5 on Mangolds
(Blocking: SPN, PKD, SKND)

I − SPKND		I + SPKND	
247	S	287	S
1	P − SKND blocks	−35	P + SKND blocks
−3		11	
47	K	−31	K
71	SK	55	SK
−19		−27	
29	SPK − ND	−37	SPK + ND
137	N	77	N
−27	SN − PKD blocks	99	SN + PKD blocks
−17		61	
3	SPN − KD blocks	−13	SPN + KD blocks
13		89	
−27		−33	
3		−7	
−141	SPKN − D	151	SPKN + D

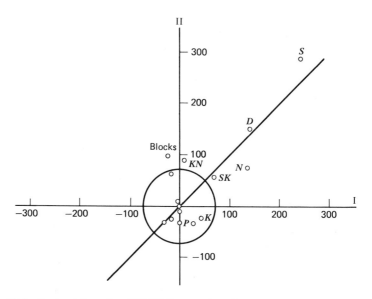

Figure 12.3 Contrasts from $I - SPKND$ (I) versus those from $I + SPKND$ (II) (Rothamsted).

12.2.4.   An Industrial 2^5 from Johnson and Leone

The 2^5 discussed by Johnson and Leone [1964, Vol. 2, page 184] has a special interest because it contains a large three-factor interaction that turned out to have a rather simple interpretation. The 32 observations coded by -8 are, in standard order, 1, 2, 0, -2, $\underline{-5}$, -3, -2, 2; 3, $\underline{5}$, 1, $\underline{8}$, -1, $\underline{2}$, -1, $\underline{5}$; -5, 1, $\underline{-4}$, -2, $\underline{-3}$, -2, $\underline{-4}$, 2; 0, -1, 0, -2, -1, -1, -3, -2. The four largest contrasts in these data are $\hat{A} = 36$, $\hat{D} = 36$, $\hat{E} = -42$, and $A\hat{D}E = -30$. These are nearly enough equal in magnitude that we code all to ± 1 and put all through the reverse Yates algorithm to see whether some pattern of response emerges:

Spec.	(0)	(1)	(2)	(3)	
ADE	-1	-1	-2	0	
DE		-1	2	0	
AE		1	0	0	
E	-1	1	0	-4	$\hat{e}$
AD		1	0	4	$\widehat{ad}$
D	1	-1	0	0	
A	1	1	-2	0	
T		-1	-2	0	

The combined impact of the four factorial contrasts is a cancellation of all effects except in two positions. These two, ad and $\hat{e}$, are equally far from the mean, one low, one high. This was not easy to see in the original data, but it does provide a simpler description of the whole set of data. The four conditions predicted to be "low" in y are the four that contain e but not a or d (e, be, ce, bce); the four predicted to be "high" are ad, abd, acd, $abcd$. These are underlined in the data listing above. There are two failures in that $\underline{\underline{c}}$ (double underline) is not supposed to be low, and acd (double underline) is not high enough.

The effects as judged by the two halves $I \pm ABCDE$ are listed in Table 12.4.

These contrasts are plotted in Figure 12.4, with a circle of the usual radius. Although most large contrasts are in the right quadrants (upper right and lower left), this fractionation must be admitted to be a failure. The rough equality of the three main effects and their 3 fi's might have been guessed from part I but is obscured by the large 2fi, DE. Both DE and ADE disappear in II!

TABLE 12.4.

I $I - ABCDE$		II $I + ABCDE$	
16	A	20	A
-4		8	
6		2	
-6		-16	
8		2	
24	$BC - ADE$	-6	$BC + ADE$
18	$-DE$	-4	
14	D	22	D
4		-8	
0		-4	
-6	$ABD - CE$	14	$ABD + CE$
-2		-8	
4		-6	
-4	$BCD - AE$	-14	$BCD + AE$
14	$ABCD - E$	-28	$ABCD + E$

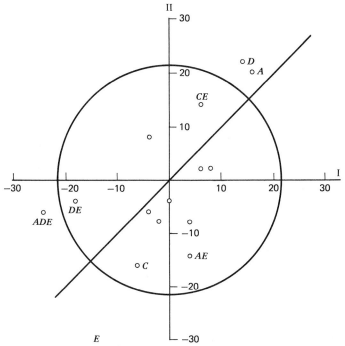

Figure 12.4 Johnson and Leone's 2^5. $I - ABCDE$ (I) versus $I + ABCDE$ (II).

212

In summary, three of our four exemplary 2^5's have survived halving, with some obscurities, but the fourth is a half failure. As we can see from the half-normal plot given in Johnson and Leone (Example 15.4, page 189), the four largest effects are only a little too large for the pattern set by the smaller 27 contrasts. This bare significance in the 2^5 is destroyed or at least seriously distorted in the halves.

12.3.   QUARTER REPLICATES: 2^{p-2}

12.3.1.   Of Resolution III, 2_{III}^{p-2}

There will be three interactions or "words" in the alias subgroup for any 2^{p-2} in addition to the identity, since effects will be aliased in sets of four. Since each letter will be present in two of these words, the average word length L will be $2p/3$. We will always want all main effects to be estimable separately; and so even if all 2fi's are negligible, as is required for a Resolution III plan, we must have all word lengths greater than or equal to 3 and hence, in symbols,

$$L \geq 3 = \frac{2p}{3}, \qquad \text{so } p \geq 4\tfrac{1}{2} \text{ or } 5.$$

In words, we cannot have a quarter replicate of Resolution III for fewer than five factors.

The 2_{III}^{5-2} is indeed realizable. We choose two 3fi's having one letter in common, say ABC and CDE, and inevitably include their product, $ABDE$, in the alias subgroup. To produce the principal block [containing treatment combination (1)] we make our confounding pattern of defining contrasts out of the negative (because odd-lettered) 3fi's, together with their necessarily positive product. Thus

$$I - ABC - CDE + ABDE$$

specifies by implication *all* the strings of four effects in our confounding pattern. The generators of the principal block are found by direct trial, requiring each to have an even number of letters in common with ABC and with CDE. Thus ab, acd, and de suffice to generate the principal block.

It is useful to write out the full aliasing for each of the contrasts that will emerge in any fractional replicate. But for the present 2^{5-2} this may suffice to repel the cautious statistician or experimenter. Thus, including only main effects and 2fi's, the expected values of the mean and of the seven standard contrasts are as follows:

$$I - ABC - CDE + ABDE$$
$$A - BC$$

$$B - AC$$
$$C - AB - DE$$
$$D - CE$$
$$E - CD$$
$$AD + BE$$
$$AE + BD$$

The experimenter must know his system very well indeed if he can ignore six 2fi's and so be able to estimate his main effects without 2fi biases.

In the 2^{5-2} given by Davies [1971, Example 10.3, pages 457 ff., 492] the alias subgroup was $I + ABE - ACD - BCDE$, and the standard deviation of single observations was known to be about 1. The data, the effects, and the aliasing are shown in Table 12.5.

TABLE 12.5.
SUMMARY OF DATA AND EFFECTS IN DAVIES' 2^{5-2}

Spec.		Yield	Alias	Total Effects (std. error = $1.0\sqrt{8}$ = 2.8)
	e	59.1		533.5
a	d	57.0	$A + BC - CD$	5.3
b		58.6	$B + AE$	23.7
ab	de	63.9	$AB + E$	0.7
c	de	67.2	$C - AD$	56.3
$a\,c$		71.6	$AC - D$	-1.1
bc	d	79.2	$BC - DE$	10.9
abc	e	76.9	$ABC + CE - BD$	-14.1

Since the standard error of the total effects must be $1.0 \times \sqrt{8}$ or 2.8, we are forced to conclude not only that B and C are large and real but also that the two aliased pairs of 2fi's are too large to be due to chance. After making up all four 2×2 tables (to show the expected yields) for BC, DE, CE, and BD, the statisticians and experimenters concluded that the conditions giving the highest yield would be bcd [Davies 1971, bottom of page 459]. Inspection of the data reveals exactly the same thing, that is, the yield at bcd is 79.2 and the next highest yield ($abce$ at 76.9) is noticeably lower. I am not implying that the standard statistical analysis should have been skipped; but when it produces results as complex in its second-order aspects as this one—clearly not expected by the experimenters—it is safer to view the whole fractional replicate as an attempt to get a broad but thin sampling of the results in

ABCDE space. Something intelligible has been learned from the fraction, namely, that factors *B* and *C* dominate the yield. If high yield is very important, then at the very least another fractional replicate should be done, and (the *P-Q* principle!) it should estimate $(BC + DE)$ and $(BD + CE)$. Since both these can be guaranteed by putting $+BCDE$ in the alias subgroup, we should arrange to separate *C* and $-AD$ by including $+ACD$ in the alias subgroup (a.s.g.) and so have, finally, $I + ACD + ABE + BCDE$. The new set will now be *odd* on *ACD* and on *ABE*. Rather than searching for all eight treatment combinations with these properties, I suggest deriving first the principal block with these defining contrasts. The new plan can then be found by multiplying all members of the principal block by *one* combination of lowercase letters that is odd on both *ACD* and *ABE*. Thus by trial I found <u>abc</u>, <u>ade</u>, and <u>be</u> as generators of the principal block, multiplied all out to get the first column of treatment combinations shown below, and then multiplied all of them by *a* (which is odd on *ACD* and *ABE*) to obtain the desired plan shown in the last column below.

Generation	Principal Block	Desired Set
	(1)	a
1	<u>abc</u>	bc
2	<u>ade</u>	de
1 × 2	bcde	abcde
3	<u>be</u>	abe
1 × 3	ace	ce
2 × 3	abd	bd
1 × 2 × 3	cd	acd

It should now be clear to the reader how to write out the aliasing pattern for the new desired set, and how to combine these results with those found earlier to separate the 2fi strings as planned. To see in full the results of combining the two 2^{5-2}'s, we note that together they make a half replicate with a.s.g. $I + ABE$ (alas), and so we still have Resolution III on factors *A*,*B*, and *E*. On the assumption that all interactions greater than the two-factor ones are negligible, the fifteen contrasts will now estimate $(A + BE), (B + AE)$, $(E + AB)$, *C*, *D*, *AC*, *AD*, *BC*, *BD*, *CD*, *CE*, and *DE*, plus three contrasts for error. The reader can see, by writing out the treatment combinations for the two quarter replicates not used, that each contains one and only one treatment combination with *B*, *C*, and *D*, all at high levels. Thus the post hoc hypothesis that *bcd* is always highest in yield can be checked after any one of the three quarters is completed.

12.3.2. Of Resolution IV, 2_{IV}^{p-2}

More commonly, the experimenter will want to see good estimates of main effects *not* aliased with any 2fi, but will be able to tolerate aliasing of main effects with 3fi's. This requires "four letter" or Resolution IV plans. Since we are still discussing quarter replicates, we now require that $L \geq 4 = 2p/3$, so that $p \geq 6$. In words, we can get a quarter replicate of Resolution IV only if at least six factors can be varied, and then in $2^{6-2} = 16$ trials.

We take $I + ABCD + CDEF + ABEF$ for our a.s.g. (any permutation of the six letters gives an equivalent plan). We write down four generators of the principal block; $\underline{ab}$, $\underline{ace}$, $\underline{ade}$, and $\underline{acf}$, say. By using as few letters as possible in each generator, and by adding one letter in each new one, we can be sure even before multiplying out that no product is like any of its predecessors. Table 12.6 gives the alias structure and the treatment combinations required. The reader is reminded that there is no correspondence between the entries in the two columns; they are printed side by side solely to save space. The last two contrasts as listed would be used as a start in estimating error.

TABLE 12.6.

ALIASING AND PLAN FOR THE 2_{IV}^{6-2}

1.	Mean + 4fi	(1)
2.	$A + $ 3fi	ab
3.	$B + $ 3fi	ace
4.	$C + $ 3fi	bce
5.	$D + $ 3fi	ade
6.	$E + $ 3fi	bde
7.	$F + $ 3fi	cd
8.	$AB + CD + EF$	$abcd$
9.	$AC + BD$	acf
10.	$AD + BC$	bcf
11.	$AE + BF$	ef
12.	$AF + BE$	$abef$
13.	$CE + DF$	$cdef$
14.	$CF + DE$	$abcdef$
15.	$ACE + BDE + ADF + BCF$	adf
16.	$ACF + BDF + ADE + BCE$	bdf

The thoughtful reader may well have been asking himself whether there is any use in a plan that mixes up 2fi's in pairs and triples. Even if such a contrast comes out experimentally to be very large, how can it be interpreted? The answer, to be expanded in Chapter 14, is that, while the sum of two or more 2fi's can be reliably decomposed only by acquiring further data, this may be only a single observation—or two, four, or eight, depending on the ratio of the magnitude of the large contrast to its standard error. With much greater risk, we sometimes jump to the conclusion that, say, the $(AC + BD)$ contrast is due largely to AC, if the effects of A and/or C are large, but those of B and D are small.

12.3.3.   Of Resolution V: 2_V^{p-2}

If a quarter replicate is to be of Resolution V, then all members of the a.s.g. must contain five or more letters. We can choose two five-letter interactions whose product is a six-letter word by allowing two letters of overlap. Thus we will have $I - ABCDE - DEFGH + ABCFGH$. Therefore the smallest 2_V^{p-2} must have $p = 8$ and so requires 64 trials. This seems quite wasteful since the number of degrees of freedom needed for eight factors is $8 \times 9/2$ or 36. We will discuss more economical alternatives in Chapter 13.

12.4.   EIGHTH REPLICATES: 2^{p-3}

There will be seven members of the a.s.g. beyond the identity, and each factor (letter) must appear four times. The average word length L must be $\geq 4p/7$. Hence for Resolution IV minimum p must be 7, and for Res. V, 9. The Res. IV plan is attainable and will be discussed below. The Res. V plan is not attainable.

Consider the a.s.g. of a 2_V^{9-3}. Start with two generators, P and Q, which must be of length 5. Their product must be of length 6. We have now consumed 16 of the $9 \times 4 = 36$ letters present in the whole a.s.g. We have, then, 20 letters left for four words, and so all must be of length 5. But this is impossible since the product of the third generator (which must be of length 5) and the two earlier generators must be of length 6.

We proceed then to a compromise plan that has Res. IV for some factors (six, in fact) and Res. V for two, all in 32 trials. We choose two 4fi's as initial generators and so get three members of the a.s.g.:

$$ABCD + ABEF + CDEF.$$

We seek a 5fi that has two letters in common with $ABCD$, two in common with $ABEF$, and hence two in common with $CDEF$, *and* that includes the two new letters to make an eight-factor plan, namely, G and H. Thus we arrive at $ACEGH$, and on multiplying this by the three members above we have the a.s.g.:

$$I + ABCD + ABEF + CDEF - ACEGH - BDEGH - BCFGH - ADFGH.$$

It is clear, is it not, that G and H are the factors all of whose 2fi's are estimable.

An experiment with these properties was reported [Daniel and Riblett 1954]. For some unknown reason the a.s.g. was not given. Let us recover it.

We pick up five generators of the principal block from Table III of the paper cited, and we then find by search *three* generators for the a.s.g.:

Generators of Principal Block		Alias Subgroup		
1. gh	1.	$\overline{ABCD}$	1×2	$CDEF$
2. $\overline{efh}$	2.	$\overline{ABEF}$		
3. $\overline{cdh}$	3.	$-\overline{ADEGH}$	1×3	$-BCEGH$
4. $\overline{bdfh}$			2×3	$-BDFGH$
5. $\overline{adf}$			$1 \times 2 \times 3$	$-ACFGH$

We see then that the 2fi's of factors A–F are aliased in strings: $AB + CD + EF, (AC + BD), (AD + BC), (AE + BF), (AF + BE), (CE + DF)$, and $(CF + DE)$. All 2fi's, including G and H, are aliased only with 3fi's.

It is worth repeating that each effect (main effect or 2fi) is estimated as precisely as if all other factors were held constant. It is worth adding that, since all 2fi and 3fi contrasts were small, they were pooled to give an estimate of random error with $31 - 8 = 23$ d.f. These estimates (two responses were measured) gave the experimenters their *first* real knowledge of the random error of the process under study. The estimates were resisted for some time, being larger than was thought acceptable, but later replicates, pooled from several separate trials, confirmed them both.

Because this was the first published multivariate (actually bivariate) fractional replicate, its data have been studied by several authors [e.g. R. Gnanadesikan and M. W. Wilk 1969], with interesting further findings and suggestions.

12.5. SIXTEENTH REPLICATES OF RESOLUTION IV: 2_{IV}^{p-4}

Since L must be at least 4, since each letter must appear 8 times, and since there are 15 interactions in the a.s.g., we have $L \geq 4 = 8p/15$, so p must be over $7\frac{1}{2}$ or 8. This fraction is actually attainable and is very well known [Box and Hunter 1961, National Bureau of Standards 1962, Hahn and Shapiro 1966, Davies 1971, Daniel 1962]. There are 28 2fi's, which are estimable only in seven strings of four each. If, as in Daniel [1962], the four 4fi's $\overline{ABCD}$, $\overline{ABEF}$, $\overline{ABGH}$, and $\overline{ACEG}$ are chosen to generate the plan, the seven strings are as follows,

$$AB + CD + EF + GH$$
$$AC + BD + EG + FH$$
$$AD + BC + EH + FG$$

$$AE + BF + CG + DH$$
$$AF + BE + CH + DG$$
$$AG + BH + CE + DF$$
$$AH + BG + CF + DE$$

Since all 15 orthogonal contrasts are used for estimation in this plan, no subdivision into blocks is possible without some further sacrifice. To divide this fraction into two blocks, we must assume away one string of 2fi's, involving (alas) each of the eight factors once. It may be more sensible to give up one factor entirely and so to revert to a 2_{IV}^{7-3}. An excellent discussion of such a plan is given in Box and Hunter [1961] with a clearly worked out example. The discussion in Davies [1971, Chapter 10, page 486] is also helpful.

12.6. RESOLUTION V FRACTIONAL REPLICATES, 2_V^{p-q}: PUBLISHED LISTS

Of the industrial experimenters with whom I have talked, a large proportion—perhaps half—have wanted to know whether their systems were "interactive." They have not usually required *all* 2fi's, but they have ordinarily insisted on getting clear estimates of some—perhaps half.

The minimum possible number, N_{min}, of trials for a Resolution V fractional replicate is $[p(p + 1) + 2]/2$, and so, for $p = 5, 6, 7, 8, 9, 10$, is 16, 22, 29, 37, 46, 56, respectively. These minima are not attainable as balanced fractions except for $p = 5$. I will indicate in the next chapter some irregular (unbalanced) fractions that have good efficiencies (although never 100%) and that come closer than the balanced fractions to the indicated minima. For the present, we stay with the regular, "orthogonal" fractions. These have, to repeat, full efficiencies for all estimates, but the price paid is sometimes high in the number of trials required.

The Res. V (and Res. IV) fractional replicates in the interesting range ($5 < p \leq 12$, $16 \leq N \leq 256$) have been listed several times. The original papers by Finney [1945] were followed by the more extended list of Brownlee, Kelly, and Loraine [1948]. Later papers by Box and Hunter [1961], by the Statistical Engineering Laboratory of the National Bureau of Standards [1962], by Hahn and Shapiro [1966], and by others have been even more extensive, more explicit, and more easily usable.

The convenient listing by Box and Hunter [1961] uses numbers to tag factors instead of letters, but this is an easy translation to make.

The National Bureau of Standards (NBS) booklet, which goes to $p = 16$, $N = 256$, does not classify the plans by resolution. However, since full a.s.g.'s are given, it is not hard to see that the smallest Res. V plans are those numbered 2.6.16, 2.7.8, 4.8.16, 4.9.16, 8.10.16, 16.11.16, 16.12.16, 64.14.32, 128.15.32, where "$q.p.b$" means a 2^{p-q} in blocks of size b. This bulletin also gives all treatment combinations explicitly, but not in standard or even alphabetical order. It is understandable that in the early 1950's it was not considered safe to transpose all these meaningless sets of letters by hand, but without an ordering of some sort it is impossible even to proofread the plans. Let us hope that someone will eventually produce a clearly printed, lexical, computer-checked printout of all these valuable plans.

Another useful compendium, which includes plans for factors at more than two levels, is *Catalog and Computer Program for the Design and Analysis of Orthogonal Symmetric and Asymmetric Fractional Factorial Experiments* by G. J. Hahn and S. S. Shapiro [1966]. This work, as well as that of Addelman and Kempthorne [1961] on orthogonal main effect plans, will be discussed in detail in Chapter 13.

12.7. FRACTIONAL REPLICATES IN THE 3^p SERIES

The authors of Davies [1971, Section 10.8, page 475] give a first-rate description of the fractionation of the 3^p factorial plans. As they write, "Fractional replication is not as satisfactory in 3^n design as in 2^n design; relatively large experiments are required to free the 2fi's even when dealing with as few as four factors." The principal reason for this unsatisfactoriness is the relatively large number of degrees of freedom in the 2fi's for the 3^p series, which is obviously four times that for the 2^p. To this increased gross requirement must be added the combinatorial restrictions, which increase still further the required number of observations.

The classic NBS document on fractions of the 3^p series by Connor and Zelen [1959] gives the best Resolution V plans. Defining as above the "degree of freedom efficiency" (d.f.e.) as the ratio of the number of degrees of freedom required for estimation of main effects and 2fi's to the number of observations specified, we have, for the smallest Res. V. plans in this document, the following:

NBS No.	p	q	N	d.f.e.
3.5.3	5	1	81	0.625
3.6.9	6	1	243	.297
9.7.9	7	2	243	.405
27.8.9	8	3	243	.529
81.9.9	9	4	243	.670
243.10.81	10	5	243	.827

If 243 trials are feasible, we may as well vary up to 10 factors. I have not seen a 243-trial experiment in my own work, but one is described in Kempthorne [1952, page 426]. The d.f.e.'s quoted above, although they come from very large plans, are nevertheless comparable with those for the 2^p series in the same range of p. (The smallest 2_V^{10-q} is of size 128 and so has a d.f.e. of 0.433.) Unless an experiment requires a long waiting time, it will usually be more economical to carry through a Res. IV plan and then to distentangle the suspiciously large contrasts measuring strings of 2fi's, as is recommended for the 2^{p-q} in Chapter 14.

When all the factors have continuous levels, the response surface methods of Box, Youle, [1954, 1955, 1957, 1958] and Hunter in [Chew, Ed. 1958, pages 138–190] and of their successors are far superior and require in general fewer trials. This is due to the reduction from 4 d.f. for each 2fi to one d.f. for the cross-product term since only that one makes sense for continuous variables.

CHAPTER 13

Incomplete Factorials

13.1.   INTRODUCTION AND JUSTIFICATION

We have praised factorial designs and their balanced fractions at great length. We have seen that, in addition to their maximum precision and minimum bias, the larger of these plans permit *study*, so that defects, even failure of some basic assumption, can be spotted. These advantages give the fractional replicates—and the full factorials—their priority when new experimental systems are being investigated, when precision is not known, and when occasional bad values are likely.

But some experimenters will feel that none of these restrictions applies to their present work. They are familiar with the system; it has constant, even known, precision; bad values are rare and are detectible by technical methods. There is, then, no great need to allow 10–30 d.f. for free-standing data analysis. All that is needed, the experimenter may insist, is a plan that will permit good estimates of the parameters in which he is interested, with perhaps some indication of their stability over a range of variation of some other experimental conditions.

In the terms of Chapters 11 and 12, there are some multifactor situations in which the experimenter is justified in demanding higher *saturation* of the design with parameter estimates—higher degree of freedom efficiency, then. The idea of resolution still holds some water, although not quite as much as earlier. The experimenter may want *some* interactions to be estimable, but not all. He may even be willing to state just which 2fi's he is quite sure are negligible, which he is doubtful about, and which he must have quite clear estimates of. The plan he wants, then, is of Resolutions III, IV, and V for different factors and factor combinations.

I have not been able to produce a logical outline for organizing this complex of situations. Hence I will simply start with plans of pure Res. III, proceed to those of Res. IV, then to mixed Res. III and IV, and so vaguely on.

The earliest work in this area, that of Plackett and Burman [1946] remains a classic, unsurpassed when a large number of factors (9–99!) are to be varied *and* when additivity of effects of all factors is assured. The penalty for the extreme saturation of these plans lies in the heavy aliasing of every main effect with 2fi's. Some lightening of this burden is offered later.

The very thorough atlas of "orthogonal main effect" plans, abbreviated in this chapter as OME plans, by Addelman and Kempthorne [1961] can hardly be praised too highly. The plans may, however, require augmentation in one respect. The atlas does not show the 2fi aliases of each main effect estimate. This is no omission if the assumption of *no* 2fi's is entirely correct. But if the experimenter's assumption of total additivity of effects is only, say, 90% secure, so that on average one 2fi in 10 is large even though all were assumed to be 0, then it will be well to know where these rare 2fi's may show, and with what main effect each is aliased. I will show how to expose these aliases, at least for the simpler plans.

Much work has been done on irregular parts of the $2^n 3^m$ series, as well as on the pure 3^m series. Some of this work, especially that of B. H. Margolin [1968], will be summarized and recommended. Rather unexpectedly, there is often a limit in that some saturated plans are of poor efficiency, but equally unexpectedly, this situation can often be improved by the addition of a small number of trials. The work of A. T. Hoke [1974] is quoted in this connection.

Very compact $2^n 3^m$ combinations in less than 20 trials, with some 2fi's estimable, are often useful for the experimenter described above, and are dangerous only if he is wrong in his judgments about negligible 2fi's. S. Webb [1965] has produced an atlas of such plans that is bound to be widely used.

13.2. PLACKETT-BURMAN DESIGNS

These plans [Plackett and Burman 1946] are described primarily to warn experimenters of their sensitivity to the multiple assumption about total additivity of effects of factors. We use the smaller plan for 11 factors in 12 trials shown in Table 13.1. Although we may take comfort from the 100% efficiency of all 11 main effect estimates, we must remember that we have "assumed away" $11 \times 10/2$ or 55 2fi's.

TABLE 13.1.
PLACKETT-BURMAN PLAN: $2^{11}//12$

Trial	A	B	C	D	E	F	G	H	J	K	L
1	0	0	0	0	0	0	0	0	0	0	0
2	0	1	1	0	1	1	1	0	0	0	1
3	1	0	1	1	0	1	1	1	0	0	0
4	0	1	0	1	1	0	1	1	1	0	0
5	0	0	1	0	1	1	0	1	1	1	0
6	0	0	0	1	0	1	1	0	1	1	1
7	1	0	0	0	1	0	1	1	0	1	1
8	1	1	0	0	0	1	0	1	1	0	1
9	1	1	1	0	0	0	1	0	1	1	0
10	0	1	1	1	0	0	0	1	0	1	1
11	1	0	1	1	1	0	0	0	1	0	1
12	1	1	0	1	1	1	0	0	0	1	0

In a 2_{III}^{p-q} the consequence of erroneously supposing a 2fi to be 0 is that *one* contrast will have a serious bias. In the Plackett-Burman (P-B) plans it turns out that all the 2fi's not involving factor P are aliased with the P-contrast. There is, then, a string of 45 2fi's behind each main effect contrast, and each 2fi appears in 9 of the 11 contrasts. The aliasing is less drastic if the plan is used for fewer than 11 factors, but each 2fi appears in more than one string of aliases.

The P-B plans are advantageous in the sense that each 2fi has coefficient $\pm\frac{1}{3}$ (not ± 1) when it appears. The pattern of signs is different for each 2fi and is identifiable for each main effect contrast. Thus, for the 11-factor plan $(A \cdots L, I$ omitted), the 2fi AB appears positive in F, J, and K, and negative elsewhere. If then all 11 contrasts were of about the same nonzero magnitude, but with signs in the pattern $- - - - - + - - + + -$, I for one would say, "This is the 2fi AB." But this pattern will not generally be clear, due to the presence of other effects and of random error. The highly irregular pattern of signs for each main effect contrast (15 pluses, 30 minuses) is not, I think, worth publishing.

As Box and Wilson showed long ago in their pioneering paper [1951], any Res. III plan for factors at two levels can be complemented by an equal plan, with all levels replaced by their alternative versions, to give a Res. IV plan. Such a scheme, with 24 trials then, seems permissible when the experimenter is quite sure that very few 2fi's are large. The prospect of disentangling a number of long strings of 2fi's, although ameliorated somewhat by the known patterns, is not enticing.

13.3. SEVERAL FACTORS AT SEVERAL LEVELS:
$2^n 3^m_{\mathrm{III, IV, V}}$

The work of Addelman and Kempthorne [1961, 1962a, b, and Appendix 13.A of this book] on OME plans is valuable as a reference if one is considering experimental work with more than, say, four factors, of which one or more are at three or more levels. The introductory 121 pages of the larger work [1961] are well worth scanning. The listing of basic plans [1961, pages 123–138] is essential. The 26 basic plans ($N = 8$–81) cover a very wide range of situations. We need not repeat the substance of the many excellent papers showing how these plans were derived. Rather, we are concerned with their use and analysis.

The OME plans to be discussed do not have alias structures as neat as those of the fractional replicates—in which each 2fi appears but once—but they are much less muddled than the P-B plans. We take Basic Plan 2 [1961, page 139] to show a tolerably simple way to find all 2fi aliases for each main effect, and we choose the most difficult case, which is for a $3^1 2^4//8$. (I have reversed the order of the factors to conform to the quoted text.)

Table 13.2 gives the plan as transcribed from Addelman and Kempthorne (henceforth A-K).

Only the contrast for A_Q is unexpected. Tables are given in A-K [1961, pages 106–119] for all other cases in which there are unequal numbers of appearances of levels of factors. Our example is shown in the second panel of their Table 18 on page 113.

Fourteen separate 2fi terms have been suppressed in the main effect parametrization. We sort these out in the third panel at the top of Table 13.2, which is labeled X_2 and is called the *confounding matrix*. As always, the set of entries in each column of X_2 is found as the product of the two corresponding numbers in the main effect columns. I find it convenient to transcribe each vector (e.g., that for A_L) onto the edge of a 3×5 index card, and then to move this set up to each other main effect set in turn. I can then write down the eight entries in each 2fi column in a few seconds without eyestrain. The marked index card should be kept for the next operation.

The X_2 matrix shows how each 2fi enters each observation. We need to know also how each 2fi enters each main effect estimate. This is particularly easy to determine when X_1, the *matrix of independent variables*, is orthogonal, as it is here, because then the direct estimates are the least squares ones. Each coefficient in the diagonal set below X_1 is just the sum of squares of the entries in X_1. The *aliases* which come with each main effect are also easy to write down. One simply puts the vertical pattern for each main effect, say A_L, namely, $-1, -1, 0, 0, 1, 1, 0, 0$, adjacent to each of the nine columns

TABLE 13.2.
THE ORTHOGONAL MAIN EFFECT $3^1 2^4 // 8$ WITH ITS 2FI ALIASES

Group headers: **X_1** spans the columns B, C; **X_2** spans the columns A_QC, A_QD.

Trial	A	B	C	D	E	μ	A_L	A_Q	B	C	D	E	A_LB+A_LD	A_QB	A_LC+A_LE	A_QC	A_QD	A_QE	$BC+DE$	$BD+CE$	$BE+CD$
1	0	0	0	0	0	1	−1	1	−1	−1	−1	−1	1	−1	1	−1	−1	−1	1	1	1
2	0	1	1	1	1	1	−1	1	1	1	1	1	−1	1	−1	1	1	1	1	1	1
3	1	0	0	1	1	1	0	−1	−1	−1	1	1	0	1	0	1	−1	−1	1	−1	−1
4	1	1	1	0	0	1	0	−1	1	1	−1	−1	0	−1	0	−1	1	1	1	−1	−1
5	2	0	1	0	1	1	1	1	−1	1	−1	1	−1	−1	1	1	−1	1	−1	1	−1
6	2	1	0	1	0	1	1	1	1	−1	1	−1	1	1	−1	−1	1	−1	−1	1	−1
7	1	0	1	1	0	1	0	−1	−1	1	1	−1	0	1	0	−1	−1	1	−1	−1	1
8	1	1	0	0	1	1	0	−1	1	−1	−1	1	0	−1	0	1	1	−1	−1	−1	1
Aliases:							4	8	8	8	8	8	4	8	4	8	8	8	8	8	8
Standard contrasts ÷ 8:							$\frac{1}{2}(C_0-BC_0)$	$-B_0$	A_0	$-AC_0$	$-AB_0$	ABC_0	$\frac{1}{2}(AC_0-ABC_0)$	$-AB_0$	$-\frac{1}{2}(A_0-AB_0)$	ABC_0	A_0	$-AC_0$	$-C_0$	$-B_0$	BC_0

(Columns A–E under "Basic Plan 2.")

of X_2 and so forms the inner product to make one entry in the row of aliases, here 0, 0, 0, −4, 0, −4, in about a minute. The first line of numbers in the alias matrix is, then;

$$E\{\hat{A}_L\} = A_L - BC - DE - BE - CD$$

Since five pairs of columns had identical entries (or their negatives), I have further shrunk the confounding and alias matrices in the column headings.

This rather blind procedure produces all seven lines of the alias matrix. The reader who requires a little more insight into the situation may want to look at the third set of panels in the table. We have here identified each of the main effect vectors with its older, more familiar name as we have used these names since Chapter 4, where they were first shown in Yates's table of signs. They are now given the subscript $_0$ to distinguish them from their current designations at the top of the table. The six contrasts $-B_0$ to ABC_0 are easy to identify, but the odd couple for A_L needs to be explained. The simplest way to make the identification is by putting the four signed "observations" for A_L through Yates's algorithm. Thus we have

(0)	(1)	(2)	(3)	
−1	−2	−2	0	
−1	0	2	0	
0	2	0	0	
0	0	0	0	
1	0	2	4	C_0
1	0	−2	0	
0	0	0	−4	BC_0
0	0	0	0	

We can now see, without a cell-by-cell check, which 2fi goes with each main effect. For example, C_0 and $-BC_0$ appear in the 2fi columns only for $(BC + DE)$ and for $-(BE + CD)$ so there are the aliases for A_L.

After such a $3^1 2^4 // 8$ is completed, one should of course put the eight results through Yates's algorithm. The "Standard contrasts" section of Table 13.2 can then be used to relabel the effects with a simple calculation, $\frac{1}{2}(C_0 - BC_0)$, added to find $4\hat{A}_L$.

The alias structure of this plan is worth looking at even if one is quite sure that no 2fi's are present. It makes explicit just what has been assumed away, and just where each 2fi might turn up. The reader will notice that no main effect is aliased with any of its own 2fi's, and that only two pairs of 2fi's, namely, the $A_L X$ components, appear in more than one main effect estimate.

Margolin [1968] uses the same example (and gives the same results!) but with a different assignment of letters to factors. Remembering that my A, B, C, D, E are his R, A, E, F, G, respectively, one can see the identity of his aliases (on his page 567) with those of Table 13.2.

Since we have handled the full $3^1 2^4$ from Basic Plan 2, it is now straightforward to see the aliasing if fewer factors are chosen from the same plan. If we drop E, making the plan a $3^1 2^3$, we simply drop all 2fi's containing E.

13.4.   AUGMENTATION OF THE $3^1 2^4//8$

When the experimenter finds one or two large effects, all others being of lesser magnitude, there is not much risk in interpreting the results as simple additive effects of the factors. But when there is no such clear separation, and when in addition the error standard deviation is not known, I would strongly advise augmenting the 8 trials by another 8, which use the opposite levels of all the two-level factors, and which interchange the levels of A as follows:

First Set	Second Set
0	1
1	0
1	2
2	1

Table 13.3 shows the new plan, labeled Plan 2_{II}, along with the corresponding X_1 and X_2 matrices and, below, the new alias matrix. By adding and subtracting the pairs of estimates for $A_L \cdots E$ (the P-Q principle), we resolve some of the aliasing. Summarizing, we can make 12 estimates from the 16 trials, namely,

$$A_L - BC - DE \qquad A_L B - 2A_Q C$$
$$A_Q \qquad\qquad\qquad A_L C - 2A_Q B$$
$$B - A_L C \qquad\qquad A_L D - 2A_Q E$$
$$C - A_L B \qquad\qquad A_L E - 2A_Q D$$
$$D - A_L E \qquad\qquad BD + CE$$
$$E - A_L D \qquad\qquad BE + CD$$

Although considerable ambiguity remains, we have been able to remove two 2fi's from each main effect string, and that accomplishment will be judged worthwhile by some. Technical identification of one of a pair of 2fi's may now be undertaken with less risk.

TABLE 13.3.

AN AUGMENTING $3^1 2^4$ WITH ITS 2fi ALIASES

Trial	Plan 2_{II} A	B	C	D	E	μ	A_L	A_Q	X_1 B	C	D	E	$A_LB - A_LD$	A_QB	$A_LC - A_LE$	X_2 A_QC	A_QD	A_QE	$BC+DE$	$BD+CE$	$BE+CD$
1	1	1	1	0	0	1	0	−1	1	−1	1	1	0	−1	0	−1	−1	−1	1	1	1
2	1	0	0	0	1	1	0	−1	−1	−1	−1	−1	0	1	0	1	1	1	1	1	1
3	0	1	0	0	1	1	−1	1	1	1	−1	−1	−1	1	−1	−1	−1	1	1	−1	−1
4	0	0	0	1	1	1	−1	1	−1	1	−1	−1	1	−1	1	1	1	1	−1	−1	−1
5	1	1	0	1	0	1	0	−1	1	−1	−1	−1	0	−1	0	−1	−1	−1	−1	1	−1
6	1	0	1	0	1	1	0	−1	−1	1	1	1	1	1	−1	−1	−1	−1	−1	1	1
7	2	1	0	1	0	1	1	1	1	−1	1	1	1	1	−1	−1	−1	−1	−1	−1	1
8	2	0	1	1	0	1	1	1	−1	1	1	−1	−1	−1	1	1	1	1	−1	−1	1
Aliases:							4	8	8	8	8	8	−4 / 4	−8	−4 / 4	−8	−8	−8	−4	−8	4
Standard contrasts ÷ 8:							$\frac{1}{2}(C_0 + BC_0)$	B_0	$-A_0$	$-AC_0$	AB_0	ABC_0	$-\frac{1}{2}(AC_0 + ABC_0)$	$-AB_0$	$\frac{1}{2}(A_0 + AB_0)$	$-ABC_0$	A_0	$-AC_0$	$-C_0$	$-B_0$	BC_0

230

Margolin [1969] has given a lower bound (let us call it N'_{IV}) for the minimum possible number of trials for a Res. IV plan for any 2^n3^m:

$$N'_{IV} = 3(n + 2m - 1).$$

For our case, $n = 4$, $m = 1$, so that $N'_{IV} = 15$. It is discouraging to me, but perhaps a challenge to scholarly statisticians, that I have not been able to come within 50% of the stated minimum.

The experimenter who feels that he is being told more than he cares to know about aliasing and dealiasing may want to consult the valuable *Catalog and Computer Program for the Design and Analysis of Orthogonal Symmetric and Asymmetric Fractional Factorial Experiments* by G. J. Hahn and S. S. Shapiro of General Electric [1966], which permits automatic estimation of all main effects and of 2fi's where feasible in any of the A-K plans. This catalog is especially useful for the choice and analysis of the larger plans. Good programs for computing any set of 2fi aliases have been written at least twice but apparently have not been published. Hoke [1970] and Webb 1965b have developed programs (Webb's is the larger) which they may well send to the first few requesters. The experienced computing statistician will recognize that for any matrix of independent variables X_1 and for any confounding matrix X_2 the alias matrix is

$$A = (X'_1X_1)^{-1}X'_1X_2,$$

as Box and Wilson showed in their classical paper [1951]. For OME plans, $(X'_1X_1)^{-1}$ is diagonal and X'_1X_2 is just the set of inner products of main effects by 2fi that I went through slowly above for Basic Plan 2. (See Margolin [1968, pages 562 ff.] for further simplifications.)

13.5. ORTHOGONAL MAIN EFFECT PLANS: $2^n3^m//16$

Addelman and Kempthorne's Basic Plan 5 [1961, page 141, and Appendix 13.A of this book] shows us how to produce all mixed plans from the $2^{12}3^1$ to the 2^33^4, and as such is a major contribution. My only addition is to give some idea of the alias patterns for one of the more complex of these alternatives. I am indebted to B. H. Margolin for having first penetrated this thicket. By methods more general (and more economical) than my simple Table 13.2, using his concept of "non-zero-sum-column vectors," he has teased out the alias pattern for the 2^63^3 given in Table 13.4. This rather repellent array should really be viewed as a great simplification. There are 63 individual 2fi's and 96 2fi terms in the array, so there is considerable duplication, but of course nothing like the amount threatened by the P-B plans. It is left as an exercise for your friendly neighborhood statistician to design an augmenting set that follows the pattern of my 2^43^1 above—or, as I would hope, does better.

TABLE 13.4.

ALIAS PATTERN OF 2fi's FOR A $2^6 3^3$ FROM BASIC PLAN 5

$A \ldots, E$ AT TWO LEVELS; $R, S, T,$ AT THREE.

1. $A - BF + CR_Q - \frac{1}{2}DT_L - \frac{1}{2}ES_L - \frac{1}{4}R_LS_L - \frac{1}{4}R_LT_L - S_QT_Q$

2. $B - AF + CS_Q - \frac{1}{2}DR_L - \frac{1}{2}ET_L - \frac{1}{4}R_LS_L - R_QT_Q + \frac{1}{4}S_LT_L$

3. $C - DE + AR_Q + BS_Q + FT_Q - \frac{1}{4}R_LT_L - R_QT_Q - \frac{1}{4}S_LT_L$

4. $D - CE - \frac{1}{4}AT_L - \frac{1}{2}BR_L - \frac{1}{2}FS_L + \frac{1}{2}R_LT_Q + \frac{1}{2}R_QS_L + \frac{1}{2}S_QT_L$

5. $E - CD - \frac{1}{4}AS_L - \frac{1}{2}BT_L - \frac{1}{2}FR_L + \frac{1}{2}R_LS_Q + \frac{1}{2}R_QT_L + \frac{1}{2}S_LT_Q$

6. $F - AB + CT_Q - \frac{1}{2}DS_L - \frac{1}{2}ER_L - \frac{1}{4}R_QT_L - R_QS_Q - \frac{1}{4}S_LT_L$

7. $R_L - BD - EF - \frac{1}{2}AS_L - \frac{1}{4}AT_L - \frac{1}{2}BS_L - \frac{1}{2}CT_L - \frac{1}{2}CS_L + ST_Q + ES_Q - EF - \frac{1}{2}FT_L + \frac{1}{2}S_QT_L + \frac{1}{2}S_LT_Q$

8. $R_Q + AC - BT_Q + \frac{1}{2}DS_L + \frac{1}{2}ET_L - FS_Q + \frac{1}{4}S_LT_L$

9. $S_L - AE - \frac{1}{4}AR_L - \frac{1}{2}BT_L - \frac{1}{2}BR_L - \frac{1}{2}CR_L - \frac{1}{2}CT_L - DF + DR_Q + ET_Q - \frac{1}{2}FT_L + \frac{1}{2}R_QT_L + \frac{1}{2}R_LT_Q$

10. $S_Q - AT_Q + BC + \frac{1}{2}DT_L + \frac{1}{2}ER_L - FR_Q + \frac{1}{4}R_LT_L$

11. $T_L - AD - \frac{1}{4}AR_L - \frac{1}{2}BS_L - BE - \frac{1}{2}CS_L - \frac{1}{2}CR_L + ER_Q + DS_Q - \frac{1}{2}FR_L - \frac{1}{2}FS_L + \frac{1}{2}R_LS_Q + \frac{1}{2}R_QS_L$

12. $T_Q - AS_Q - BR_Q + CF + \frac{1}{2}DR_L + \frac{1}{2}ES_L + \frac{1}{4}R_LS_L$

13.6. SMALL, INCOMPLETE $2^n 3^m$ PLANS ($N \leq 20$)

This section is entirely devoted to recommendations for, and warnings about, the catalog of Webb [1965], which collects about 67 plans, most of them developed by Webb himself. Some of these designs are outrageously small (4, 5, 6 trials). The reader should be warned that Webb considers his factor levels as quantitative and only permits estimation of 2fi's including three-level factors of the form $A_L B$ or $A_L B_L$ (not $A_L B_Q$ or $A_Q B_Q$).

It is hard to see why Webb calls the plans $3^n 2^m$, whereas Margolin terms them $2^n 3^m$. We are in the hands of mathematicians, who cannot be required to use a lexical or even a uniform notation.

By far the easiest way to analyze any of these irregular plans is by a standard least-squares regression program, arranged to give all effects and interactions as regression coefficients, with their standard errors, t-values, and perhaps component effects and degrees of orthogonality.

Since most of these plans are nearly saturated, there is little to be learned from the residuals from a full fitting equation. Of course if only 3 or 4 d.f. are consumed by large effects and 10–15 are left for lack of fit, the residuals will not be severely distorted and may reveal a wild value or even a localized interaction.

13.7. ESTIMATION OF SELECTED SUBSETS OF TWO-FACTOR INTERACTIONS

Addelman [1962b] has given a useful list of plans (all are fractional factorials from the 2^n to the 5^q series) which permit us to estimate *some* 2fi's. He has arranged his plans in three classes. Class One plans are for k factors ($k = 3 \cdots 62$) of which a subset of size d may interact, but only with each other. Class Two plans are for k factors of which a subset of size d interact only with each other, *and* the remaining set of size $k - d$ factors also interact with each other. Class Three plans provide estimation of all 2fi's with a specified subset of factors.

Addelman's three tables are reproduced here (with permission) as Tables 13.5, 13.6, and 13.7. As a first example of their use, suppose that the effects of 10 two-level factors are under study, and that 4 of them are thought likely to interact with each other. Table 13.5 (Addelman's Table 1) shows us that we must use 32 trials to accommodate these requirements. If we use A, B, C, D for the four interacting factors, then any of the 3-, 4-, 5-factor interactions, or even AE, BE, CE, DE, may serve as the noninteracting factors.

TABLE 13.5.

ADDELMAN'S TABLE 1: FACTOR REPRESENTATIONS FOR CLASS ONE
COMPROMISE PLANS

Number of Trials	Number of Interacting Factors	Total Number of Factors	Factor Representations
2^3	2	6	$A, B; C, AC, BC, ABC$
	3	4	$A, B, C; ABC$
2^4	2	14	$A, B; C, D$, all interactions excluding AB
	3	12	$A, B, C; D$, all interactions excluding AB, AC, and BC
	4	9	$A, B, C, D; ABC, ABD, ACD, BCD, ABCD$
	5	5	$A, B, C, D, ABCD;$
2^5	2	30	$A, B; C, D, E$, all interactions excluding AB
	3	28	$A, B, C; D, E$, all interactions excluding AB, AC, and BC
	4	25	$A, B, C, D; E, AE, BE, CE, DE$, all 3-, 4-, and 5-factor interactions
	5	21	$A, B, C, D, E;$ all 3-, 4-, and 5-factor interactions
	6	16	$A, B, C, D, E, ABCDE;$ all 3-factor interactions

TABLE 3.5 (*continued*)

Number of Trials	Number of Interacting Factors	Total Number of Factors	Factor Representations
2^6	2	62	A, B; C, D, E, F, all interactions excluding AB
	3	60	A, B, C; D, E, F, all interactions excluding AB, AC, and BC
	4	57	A, B, C, D; E, F, all interactions excluding AB, AC, AD, BC, BD, and CD
	5	53	A, B, C, D, E; F, AF, BF, CF, DF, EF, all 3-, 4-, 5-, and 6-factor interactions
	6	48	A, B, C, D, E, F; all 3-, 4-, 5-, and 6-factor interactions
	7	42	A, B, C, D, E, F, $ABCDEF$; all 3- and 4-factor interactions
	8	35	$A, B, C, D, E, F, ABCD, ABEF$; ACE, ACF, ADE, ADF, BCE, BCF, BDE, BDF, CDE, CDF, CEF, DEF, $ACDEF$, $BCDEF$, $ABCDEF$, all remaining 4-factor interactions excluding $CDEF$
3^3	2	11	A, B; C, AC, AC^2, BC, BC^2, ABC, ABC^2, AB^2C, AB^2C^2
	3	7	A, B, C; ABC, ABC^2, AB^2C, AB^2C^2
3^4	2	38	A, B; C, D, all interactions excluding AB and AB^2
	3	34	A, B, C; D, all interactions excluding AB, AB^2, AC, AC^2, BC, and BC^2
	4	28	A, B, C, D; all 3- and 4-factor interactions
	5	20	A, B, C, D, $ABCD$; ABC^2, AB^2C, AB^2C^2, ABD^2, AB^2D, AB^2D^2, ACD^2, AC^2D, AC^2D^2, BCD^2, BC^2D, BC^2D^2, ABC^2D^2, AB^2CD^2, AB^2C^2D
4^3	2	18	A, B; C, all interactions excluding AB, AB^2, and AB^3
	3	12	$A, B, C; ABC$, ABC^2, ABC^3, AB^2C, AB^2C^2, AB^2C^3, AB^3C, AB^3C^2, AB^3C^3
5^3	2	27	A, B; C, all interactions excluding AB, AB^2, AB^3, and AB^4
	3	19	A, B, C; all 3-factor interactions

234

TABLE 13.6.

ADDELMAN'S TABLE 2: FACTOR REPRESENTATIONS FOR CLASS TWO
COMPROMISE PLANS

Number of Trials	Number of Factors		Factor Representations
	First Set	Second Set	
2^5	2	5	$A, B; C, D, E, ACD, BCE$
	3	4	$A, B, C; D, E, ABC, ADE$
2^6	2	8	$A, B; C, D, E, F, ACD, BCE, AEF, BCDF$
	3	7	$A, B, C; D, E, F, ABC, ADE, BDF, CDEF$
	4	6	$A, B, C, D; E, F, ABC, BCDE, ACDF, BCEF$
	5	5	$A, B, C, D, E; F, ABC, ADE, BDEF, ACEF$
2^7	2	11	$A, B; C, D, E, F, G, ACD, BCE, ABCDE,$ $CDEF, ADEG, ABCFG$
	3	10	$A, B, C; D, E, F, G, ABC, ABDE, ACFG,$ $BCDF, DEFG, BDEG$
	4	9	$A, B, C, D; E, F, G, ABC, CEF, DEG, ADFG,$ $BCEG, BDEFG$
	5	8	$A, B, C, D, E; F, G, ABC, ADE, ACEF, BCDG,$ $ABFG, DEFG$
	6	7	$A, B, C, D, E, F; G, ABC, ADE, BDF, AEFG,$ $BCDG, ABDEG$
3^4	2	4	$A, B; C, D, ACD, BCD^2$
	3	3	$A, B, C; D, ABC, ABC^2D$

TABLE 13.7.

ADDELMAN'S TABLE 3: FACTOR REPRESENTATIONS FOR CLASS THREE
COMPROMISE PLANS

Number of Trials	Number of Interacting Factors	Total Number of Factors	Factor Representations
2^3	0	7	A, B, C, AB, AC, BC, ABC
	1	4	$A; B, C, BC$
	2	3	$A, B; C$
	3	3	$A, B, C;$
2^4	0	15	$A, B, C, D,$ all interactions
	1	8	$A; B, C, D, BC, BD, CD, BCD$
	2	5	$A, B; C, D, CD$
	3	5	$A, B, C; D, ABCD$
	4	5	$A, B, C, D; ABCD$
	5	5	$A, B, C, D, ABCD;$
2^5	0	31	$A, B, C, D, E,$ all interactions
	1	16	$A; B, C, D, E,$ all interactions not containing A

TABLE 13.7 (*continued*)

Number of Trials	Number of Interacting Factors	Total Number of Factors	Factor Representations
	2	9	$A, B; C, D, E, CD, CE, DE, CDE$
	3	9	$A, B, C; D, E, DE, ABCD, ABCE, ABCDE$
	4	7	$A, B, C, D; E, ABCD, ABCDE$
	5	6	$A, B, C, D, E; ABCDE$
	6	6	$A, B, C, D, E, ABCDE;$
2^6	0	63	A, B, C, D, E, F, all interactions
	1	32	$A; B, C, D, E, F$, all interactions not containing A
	2	17	$A, B; C, D, E, F$, all interactions not containing A or B
	3	17	$A, B, C; D, E, F, DE, DF, EF, DEF,$ $ABCD, ABCE, ABCF, ABCDE, ABCDF,$ $ABCEF, ABCDEF$
	4	11	$A, B, C, D; E, F, EF, ABCD, ABCDE,$ $ABCDF, ABCDEF$
	5	10	$A, B, C, D, E; F, ABCF, ADEF, BCDE,$ $BCDEF$
	6	9	$A, B, C, D, E, F; ABCD, ABEF, CDEF$
	7	8	$A, B, C, D, E, F, ABCD; ABEF$
	8	8	$A, B, C, D, E, F, ABCD, ABEF;$
3^3	0	13	$A, B, C, AB, AB^2, AC, AC^2, BC, BC^2, ABC,$ ABC^2, AB^2C, AB^2C^2
	1	5	$A; B, C, BC, BC^2$
	2	3	$A, B; C$
	3	3	$A, B, C;$
3^4	0	40	A, B, C, D, all interactions
	1	14	$A; B, C, D$, all interactions not containing A
	2	6	$A, B; C, D, CD, CD^2$
	3	6	$A, B, C; D, ABCD, ABCD^2$
	4	5	$A, B, C, D; ABCD$
	5	5	$A, B, C, D, ABCD;$
4^3	0	21	A, B, C, all interactions
	1	6	$A; B, C, BC, BC^2, BC^3$
	2	3	$A, B; C$
	3	3	$A, B, C;$
5^3	0	31	A, B, C, all interactions
	1	7	$A; B, C$, all interactions not containing A
	2	3	$A, B; C$
	3	3	$A, B, C;$

For an example from Table 13.6 suppose that 7 factors are to be varied, that A, B, C interact, and that D, E, F, G interact but only among themselves. Line 2 of the table tells us that again we require a 2^5, that A, B, C are to be used for the first subset, and that D, E, ABC, ADE must be used for the other subset. The reader must see by now that, whereas D and E can be used directly, F and G are to be represented by ABC and ADE, so that the six 2fi's in the second set are to be identified with the contrasts in the 2^5 as follows:

$$DE = DE,$$
$$DF = D \times ABC = ABCD,$$
$$DG = D \times ADE = AE,$$
$$EF = E \times ABC = ABCE,$$
$$EG = E \times ADE = AD,$$
$$FG = ABC \times ADE = BCDE.$$

These six are all easily identifiable in the 2^5, all are mutually orthogonal, and none involves main effects or 2fi's among A, B, and C.

APPENDIX 13.A

ORTHOGONAL MAIN EFFECT PLANS FROM ADDELMAN AND KEMPTHORNE [1961]

BASIC PLAN 1: 2^3; 4 trials

```
123
000
011
101
110
```

BASIC PLAN 2: 4; 3; 2^7; 8 trials

```
*   *   1234567
0   0   0000000
0   0   0001111
1   1   0110011
1   1   0111100
2   2   1010101
2   2   1011010
3   1   1100110
3   1   1101001
    *-1,2,3
```

Basic Plan 5: 4^5; 3^5; 2^{15}; 16 trials

```
12345   12345   00000 00001 11111
*****   *****   12345 67890 12345
00000   00000   00000 00000 00000
01123   01121   00001 10111 01110
02231   02211   00010 11011 10011
03312   01112   00011 01100 11101
10111   10111   01100 00110 11011
11032   11012   01101 10001 10101
12320   12120   01110 11101 01000
13203   11201   01111 01010 00110
20222   20222   10100 01011 01101
21301   21101   10101 11100 00011
22013   22011   10110 10000 11110
23130   21110   10111 00111 10000
30333   10111   11000 01101 10110
31210   11210   11001 11010 11000
32102   12102   11010 10110 00101
33021   11021   11011 00001 01011
```

```
1-000   2-000   3-000   4-111   5-111
*-123   *-456   *-789   *-012   *-345
```

Basic Plan 6: 8; 7; 6; 5; 2^8; 16 trials

```
1 1 1 1   23456789
0 0 0 0   00000000
0 0 0 0   11111111
1 1 1 1   00001111
1 1 1 1   11110000
2 2 2 2   00110011
2 2 2 2   11001100
3 3 3 3   00111100
3 3 3 3   11000011
4 4 4 4   01010101
4 4 4 4   10101010
5 5 5 1   01011010
5 5 5 1   10100101
6 6 2 2   01100110
6 6 2 2   10011001
7 3 3 3   01101001
7 3 3 3   10010110
```

238

BASIC PLAN 13: 4^9; 3^9; 2^{11}; 32 trials

123456789	123456789	00000	00001	11111	11112	22222	22	2233
*********	*********	12345	67890	12345	67890	12345	67	8901
000000000	000000000	00000	00000	00000	00000	00000	00	0000
011231111	011211111	00001	10111	01110	01101	10110	11	0000
022312222	022112222	00010	11011	10011	10110	11011	01	0000
033123333	011121111	00011	01100	11101	11011	01101	10	0000
101111032	101111012	01100	00110	11011	01100	01101	01	0011
110320123	110120121	01101	10001	10101	00001	11011	10	0011
123203210	121201210	01110	11101	01000	11010	10110	00	0011
132032301	112012101	01111	01010	00110	10111	00000	11	0011
202223102	202221102	10100	01011	01101	11001	10001	01	0101
213012013	211012011	10101	11100	00011	10100	00111	10	0101
220131320	220111120	10110	10000	11110	01111	01010	00	0101
231300231	211100211	10111	00111	10000	00010	11100	11	0101
303332130	101112110	11000	01101	10110	10101	11100	00	0110
312103021	112101021	11001	11010	11000	11000	01010	11	0110
321020312	121020112	11010	10110	00101	00011	00111	01	0110
330211203	110211201	11011	00001	01011	01110	10001	10	0110
002130213	002110211	00000	01010	11110	00010	10111	10	1111
013301302	011101102	00001	11101	10000	01111	00001	01	1111
020222031	020222011	00010	10001	01101	10100	01100	11	1111
031013120	011011120	00011	00110	00011	11001	11010	00	1111
103021221	101021221	01100	01100	01011	01110	11010	11	1100
112210330	112210110	01101	11011	01011	00011	01100	00	1100
121333003	121111001	01110	10111	10110	11000	00001	10	1100
130102112	110102112	01111	00000	11000	10101	10111	01	1100
200313311	200111111	10100	00001	10011	11011	00110	11	1010
211122200	211122200	10101	10110	11101	10110	10000	00	1010
222001111	222001111	10110	11010	00000	01101	11101	10	1010
233230022	211210022	10111	01101	01110	00000	01011	01	1010
301202323	101202121	11000	00111	01000	10111	01011	10	1001
310033232	110011212	11001	10000	00110	11010	11101	01	1001
323110101	121110101	11010	11100	11011	00001	10000	11	1001
332321010	112121010	11011	01011	10101	01100	00110	00	1001

1-000	2-000	3-000	4-111	5-111	6-111	7-122	8-222	9-222
*-123	*-456	*-789	*-012	*-345	*-678	*-901	*-234	*-567

239

CHAPTER 14

Sequences of Fractional Replicates

14.1. INTRODUCTION

Fractional replicates are done by those who do them because they are economical. They are not done by those who do not do them because they are risky. The risk lies in the ambiguity caused by their inevitable aliasing. Experimenters sometimes feel that they can safely judge which 2fi's are negligible and hence that they can securely interpret the strings of main effects and 2fi's which the fractional replicates produce. They are often right, and many successful fractional replicates are in the record.

It does happen, however, that some fractional replicates produce ambiguous results which the experimenter may want to resolve by further work. This chapter is devoted to proposals for *dealiasing* certain effects, if the reader will pardon the neologism. It is restricted to the augmentation of the 2^{p-q} series, although there is no doubt that similar systems can be worked out for other types of designs.

To put it at its simplest, we use the P-Q principle, sometimes extended to the P-Q-R-S principle. When $P + Q$ is known or at least estimated, we produce plans which estimate $P - Q$—or $P - Q + A + \mu$, where A and μ are already estimable—and then we combine the two. We have already used this principle in designing plans before any work is done. We now extend it to fix up data sets that have failed to produce clear results.

The second-order response surface methodology of Box, Youle, and Hunter [1954, 1955, 1957] and of Hill, and Hunter [1966] may also be viewed as a system of augmenting Resolution V 2^{p-q} designs in order to get more general and more intelligible results when all factors are continuous. The augmentations of this chapter are of a more primitive type. They are aimed at discovering *which* 2fi's are there, and so in some cases will simplify the response surface fitting that will come later. They will more frequently be used after Res. IV plans, which commonly produce strings of 2fi's as estimable quantities.

Most of the proposals made here can be found in the paper on sequences [Daniel 1962]. Perhaps they are described more clearly here.

14.2. SIMPLEST AUGMENTATIONS: THE 2_{III}^{3-1} AND THE 2_{IV}^{4-1}

The first paper on augmentations of 2^{p-q} to separate aliases was that of Davies and Hay [1950]. They required the addition of at least one more set as large as the original set. The plans proposed below are generally smaller, although cases will arise that demand equal or larger extra sets.

It must be obvious to any reader who is not opening this work to this chapter that the 2^{3-1}, $I - ABC$, yields three contrasts, (A), (B), (C), which have expected values $(A - BC)$, $(B - AC)$, and $(C - AB)$. Each main effect is aliased with the complementary 2fi with opposite sign. The expected value of each trial in the *other* 2^{3-1}, $I + ABC$, will have matching signs on each main effect and its attached 2fi. Thus we have

$$E\{a\} = \mu + A - B - C - AB - AC + BC \, (+ABC).$$

Now after the first 2^{3-1} is done, if it so happens that only the A-contrast is large, there is only one serious ambiguity to resolve. When we dare say, then, that B, C, AB, AC are all roughly 0, we can write

$$E\{a\} = \mu + A + BC.$$

If we carry out only trial a, we can use the estimate of μ from the first fraction and so get an estimate of $A + BC$:

$$A \, \hat{+} \, BC = \hat{\mu} - a.$$

We carry through the computation on the five responses because of the rather counterintuitive outcome. We compute $4(A \, \hat{+} \, BC)$ as described just above, write down the estimate $4(A \, \hat{-} \, BC)$ from the original fraction, and then add and subtract:

Step		(1)	ab	ac	bc	a	
1.	$-4\hat{\mu}$	-1	-1	-1	-1		
2.	$4a$					4	
3. = 1. + 2.	$4(A \,\hat{\mp}\, BC)$	-1	-1	-1	-1	$+4$	
4.	$4(A \,\hat{\frown}\, BC)$	-1	$+1$	$+1$	-1		
5. = $\frac{1}{2}$(3. + 4.)	$4\hat{A}$	-1			-1	$+2$	
6. = $\frac{1}{2}$(3. − 4.)	$4BC$		-1	-1		$+2$	

From steps 5 and 6 we see immediately that Var $(\hat{A})$ = Var $(\hat{BC})$ = $\frac{3}{8}$; and since the minimum variance for five observations is $\frac{1}{4}(\frac{1}{3} + \frac{1}{2})$ = $\frac{5}{24}$, we have attained an efficiency of $\frac{5}{9}$. (Henceforth in this chapter we will take σ^2 to be 1.)

It is shocking, is it not, to find that we require only three of the five observations for the unaliased estimates, even though it is pleasing that the efficiency of each estimate is $\frac{5}{9}$, based on all five observations. This sounds slightly superefficient.

The adding of a single trial to a 2^{3-1} is safe only when the system is assuredly stable, without time drift. A minimum *block* for separating A from $-BC$ will contain two trials. We are allowed to use only the within-block contrast in estimation. Although the pair can be found by trial, it helps our insight to derive it formally. Since we want a contrast that includes $A + BC$, we choose an alias subgroup that contains only *other* effects. We must include $+ABC$ since all trials not yet made are in the 2^{3-1}: $I + ABC$. We add $-AC$ to the group and so are forced to include the product $+ABC \times -AC = -B$. The trials corresponding can be found by reverse Yates algorithm, putting 1, -1, -1, 1 in the positions of the four elements of the group. This pair is of course not the only one that meets our requirements. If we had chosen $I + ABC - BC - A$, the pair b, c would emerge.

Returning to the pair a, c, we see that the contrast $(a - c)$ has expectation $2(A - C - AB + BC) = 2(A + BC)$ since we have "decided" that C and AB are negligible. We now need no correction for μ, and so we can utilize the P-Q principle to combine this estimate with that for $A - BC$:

	(1)	ab	ac	bc	a	c
$4(A \,\hat{\frown}\, BC)$	-1	1	1	-1		
$4(A \,\hat{\mp}\, BC)$					2	-2
$8\hat{A}$	-1	1	1	-1	2	-2
$8\hat{BC}$	-1	1	1	-1	-2	2

We see that Var $(\hat{A})$ = Var $(\hat{BC})$ = $\frac{3}{16}$ and hence that the efficiencies are $\frac{1}{6} \times \frac{16}{3} = \frac{8}{9}$.

We proceed to the 2_{IV}^{4-1} for the simplest case of two 2fi's aliased with each other. The reader will remember that the 2_{IV}^{4-1}: $I + ABCD$ provides seven orthogonal contrasts, one for each main effect and three for the three pairs of complementary interactions. Suppose that $(AB + CD), (A)$, and (C) "come out large" and that $(B), (D)$, and the other two interaction sums are small. Any treatment combination not in the first 2^{4-1} will do, for example,

$$E\{a\} = \mu + A - C - AB + CD \pm 11 \text{ negligible terms.}$$

Since, then, $(AB \doteq CD) = \hat{\mu} - \hat{C} + \hat{A} - a$, our single observation requires three corrections. We tabulate $8(AB \stackrel{\frown}{\mp} CD)$ from the 2^{4-1}, and then $8(AB \doteq CD) = 8\hat{\mu} - 8\hat{C} + 8\hat{A} - 8a$ as they depend on the nine pieces of data.

	(1)	ab	ac	bc	ad	bd	cd	abcd	a
$8(AB \stackrel{\frown}{\mp} CD)$	1	1	-1	-1	-1	-1	1	1	
$+8\hat{\mu}$	1	1	1	1	1	1	1	1	
$-8\hat{C}$	1	1	-1	-1	1	1	-1	-1	
$+8\hat{A}$	-1	1	1	-1	1	-1	-1	1	
$-8a$									-8
$16\hat{A}B$	2	4	0	-2	2	0	0	2	-8
$16\hat{C}D$	0	-2	-2	0	-4	-2	2	0	+8
$8\hat{A}B$	1	2	0	-1	1	0	0	1	-4
$8\hat{C}D$	0	-1	-1	0	-2	-1	2	0	+4

Var $(\hat{A}B)$ = Var $(\hat{C}D)$ = $\frac{24}{64} = \frac{3}{8}$; Min Var (9 obs.) = $\frac{9}{80}$. Hence efficiencies = $\frac{3}{10}$.

The reader will find, if he carries the AB-contrast through Yates's forward algorithm, that A and C have coefficients 0, while only AB has coefficient 8. There are 11 other nonzero multipliers, but these are all effects that we have assumed negligible.

Our efficiency, $\frac{3}{10}$, is not dazzling, but I do not find it disturbing either. It will be even smaller for single augmentations of larger plans. I view it as satisfactorily far above 0, rather than as disastrously far below 1.

We now choose a pair of trials for the same case. We take care that both are at the same levels of A and of C to avoid variance-increasing corrections. Our alias subgroup will then contain $-ABCD$, $+A$, and, say, $-C$ and so will be:

$$I - \underline{ABCD} + \underline{A} - \underline{C} - AC - BCD - ABD + BD.$$

Now our desired contrast will estimate

$$B - ACD + \underline{AB} - BC - ABC - \underline{CD} - AD + D \doteq AB - CD.$$

I find it simplest to pick the two trial specifications by direct search, but of course reverse Yates on the eight signed members of the alias subgroup will produce the same result, namely, *abd* and *a*.

If we designate the original contrast for $8(AB + CD)$ by y_1 (with variance 8), and the new contrast from the pair by y_2 (with variance 2), we obtain:

$$16\hat{AB} = y_1 + 4y_2; \qquad 16\hat{CD} = y_1 - 4y_2,$$

each with variance $\frac{5}{32}$ and so with efficiency 0.64.

It is not difficult to produce a set of *four* trials if they are necessary to get more precision in separating AB from CD, when A and C (only) are large. We use the alias subgroup $I - ABCD - AC + BD$ and acquire immediately the set: *a*, *c*, *abd*, *bcd*. But since we now have three contrasts (among the four augmenting trials) at our disposal, we consider what use to make of the other two, which are

$$A - C + ABD - BCD \rightarrow A - C \qquad \text{and} \qquad B - ABC + D - ACD.$$

We should use the first to improve our precision of estimation of A and C, and the second as a degree of freedom for error. Designate the original estimate of A as, A_1. The new estimate, A_2, will be $(A \doteq C)$ from the new set of four, *plus* $\hat{C}_1$ estimated from the main experiment. We will combine these two estimates by weighting them inversely as their variances. Variance $(\hat{A}_1) = \frac{1}{8}$, and $\text{Var}(\hat{A}_2) = \text{Var}(A \doteq C)_2 + \text{Var}(\hat{C}_1) = \frac{1}{4} + \frac{1}{8} = \frac{3}{8}$, whence, as we mathematicians say,

$$\hat{A} = \frac{3A_1 + A_2}{4},$$

$$\text{Var}(\hat{A}) = \frac{9\,\text{Var}(\hat{A}_1) + \text{Var}(\hat{A}_2)}{16} = \frac{9}{8} + \frac{3}{8} = \frac{3}{32}$$

$$\text{Eff.}(\hat{A}) = \frac{1}{12} \times \frac{32}{3} = \frac{8}{9}.$$

The same system is used to improve our earlier C-estimate.

If now we suppose that *two* pairs of 2fi's require disentanglement, we find that the only augmenting set of four that meets our requirements puts all four 2fi's into a single string and so cannot be used. We might hope to use one of Addelman's plans for estimating all main effects and 2fi's in four factors in 12 runs (a $2^4//12$, then), but this fails since his plans do not include a 2_{iV}^{4-1}. I see no satisfying alternative to the full set of eight, $I - ABCD$, if two 2fi interaction strings appear large after the first half replicate.

14.3. AUGMENTING THE 2_{IV}^{8-4}

14.3.1. To Separate a Single Pair of Two-Factor Interactions

Since there is nothing new in this case, the separation design is left as an exercise. Slightly better use of the remaining contrasts (when *four* augmenting trials are made) is available because we now get an estimate of C and one of A separated from each other. Each is used to decrease the variance of the weighted estimate of a main effect.

14.3.2. To Separate Four Members of a Single String of Two-Factor Interactions

If four main effects, say A, C, E, G, and a single string of 2fi's, say $AB + CD + EF + GH$, all appeared large in a 2_{IV}^{8-4}, the experimenter might well wish to isolate all four 2fi components. Although it is possible to name *three* treatment combinations that will permit estimation of the four 2fi's [Daniel 1962], their variances are so large as to be generally unacceptable. Even the use of three pairs of observations gives poor efficiencies. (The three pairs given in the 1962 paper [page 409, line 11] are erroneous. The corrections given later [Daniel 1963] are rather opaque.) The simplest pairs are $df - bh$ (to estimate the ordered string with signs $+ \; - \; - \; +$), $dh - bf$ (to get $+ \; - \; + \; -$), and $fh - bd$ (for $+ \; + \; - \; -$).

As we look at these six conditions, we note that all are in the $2_{IV}^{4-1} : I + BDFH$. We add the missing two, namely, (1) and $bdfh$, to form the full "even" half replicate on B, D, F, and H, knowing full well that the eight are really a 2^{8-5}. I have chosen the low levels of A, C, E, and G so that all treatments are at the same levels of the four influential factors. In this way I avoid corrections for main effects. The alias subgroup for our eight trials is then:

$$I + BDFH - A - C - E - G + AC$$
$$+ \; AE + AG + CE + CG + EG \; plus \; \text{5fi's.}$$

We extract directly the four contrasts containing our 2fi components by multiplication of each member of the alias subgroup in turn:

$$y_1 = \underline{AB} - B + BC + BE + BG \pm \; \geq \; 3\text{fi,}$$
$$y_2 = \underline{CD} - D + AD + DE \pm \; \geq \; \text{fi,}$$
$$y_3 = \underline{EF} - F + AF + CF \pm \; 3\text{fi,}$$
$$y_4 = \underline{GH} - H + AH + EH \pm \; 3\text{fi.}$$

So our four 2fi's are separated from each other by this "interaction extraction fraction" (D. W. Behnken) and are aliased only with clouds of effects we have already judged to be negligible. The remaining three contrasts in the 2^{8-5}, which may be called $BD + FH$, $BF + DH$, and $BH + DF$, can be used as error estimates.

Again we can attain a somewhat smaller variance for the 2fi estimates by combining each one given above with the estimate of their sum (call it y_5) from the original 2^{8-4}:

$$AB_1 = y_1, \text{Var}(AB_1) = \tfrac{1}{8} = \tfrac{2}{16}$$

$$AB_2 = y_5 - y_2 - y_3 - y_4, \text{Var}(AB_2) = \tfrac{1}{16} + \tfrac{1}{8} + \tfrac{1}{8} + \tfrac{1}{8} = \tfrac{7}{16};$$

$$AB = \frac{7AB_1 + 2AB_2}{9}, \text{Var}(\widehat{AB}) = \frac{49 \times \tfrac{1}{8} + 4 \times \tfrac{7}{16}}{81} = \tfrac{7}{72}.$$

Eff. $(\widehat{AB}) = \tfrac{1}{24} \times \tfrac{72}{7} = \tfrac{3}{7}$.

Although algebraic formulae can be worked out for all cases· it seems to me simpler to make a computation like the above, always using the $P\text{-}Q$ principle when feasible, and otherwise always weighting each (orthogonal) estimate inversely as its variance.

14.3.3. To Separate All Seven Two-Factor Interactions with One Factor

The separation of all 2fi's with one factor is analogous to the preceding case, although I did not know this in 1962. The seven 2fi's with A, for example, appear in different alias strings, so we look for a set of trials from which we can estimate $AB - CD - EF - GH$, and also $AC - BD - EG - FH$, and so on down to $AH - BG - CF - DE$. Clearly the alias subgroup must contain $-ABCD$, $-ABEF$, $-ABGH$, and $-ACEG$, and equally clearly we require one more generator since we hope to do only $8 = 2^{8-5}$ trials. We use $-A$ as the last generator, so that all trials will be at low A. The alias subgroup now contains 32 terms, but they are easy to view overall. The first 16 members are:

123	$-ABCDEFGH$	34	$+BCEH$
1	$-ABCD$	24	$+BCFG$
2	$-ABEF$	14	$+BDEG$
3	$-ABGH$	1234	$+BDFH$
4	$-ACEG$	12	$+CDEF$
234	$-ACFH$	13	$+CDGH$
134	$-ADEH$	23	$+EFGH$
124	$-ADFG$		I

and the second 16 members are the products of these with $-A$.

The corresponding trial specifications or treatment combinations are:

bcdefgh	*ceg*
bcd	*cfh*
bef	*deh*
bgh	*dfg*

These are seen to be the lowercase counterparts of the first column of the alias subgroup with *a* removed.

It is by now an elementary exercise to combine each of the seven contrasts from this 2^{8-5}, after correction by a main effect if large, with the corresponding string from the original 2^{8-4}. I give an example which I hope most readers can skip. Designate the $-B$-contrast from the 2^{8-5} as y_3. Then

$$E\{y_3\} = -B + AB - (CD + EF + GH).$$

Let

$$y_1 = B_1 \cdots \text{ from the } 2^{8-4},$$
$$y_2 = AB + CD \hat{\mp} EF + GH \quad \text{from the } 2^{8-4}.$$

Then

$$AB + (CD \hat{\mp} EF + GH) = y_2$$
$$AB - (CD + EF + GH) = y_1 + y_3$$
$$\overline{\hat{AB} \qquad\qquad\qquad = \tfrac{1}{2}(y_1 + y_2 + y_3)}$$

and

$$\text{Var}(\hat{AB}) = \tfrac{1}{4}(\tfrac{1}{16} + \tfrac{1}{16} + \tfrac{1}{8}) = \tfrac{1}{16},$$

$$\text{Eff.}(\hat{AB}) = \tfrac{1}{24} \times 16 = \tfrac{2}{3}.$$

Similar operations with the remaining six contrasts from the 2^{8-5} will separate the other interactions with A.

The paper "Sequences of Two-Level Fractional Factorial Plans" by S. Addelman [1969] discusses the same problem as this chapter, at greater length and for a wider range of designs, but with less tailoring to specific cases. I start with minimal augmentations ($N = 1, 2, 4, 8$), responding to just what has been exposed earlier, and proposing only the smallest augmentations which will remove known ambiguities with acceptably small variance.

Addelman uses more general criteria. He shows what set to add to each small initial block to gain maximal numbers of estimates of parameters. His augmenting sets are usually as large as the initial block. The second block is chosen for the large number of new estimates it permits, but these are not explicitly tailored to the set of ambiguities revealed by the first set. The tailored, irregular, post hoc plans that I propose are not, as Addelman very rightly says, easily categorized and listed.

14.4. CONCLUSIONS AND APOLOGIES

I have viewed the known aliases (of second order) of an estimate as amguities which can be counted. The separation of a string (sums and differerences) of aliases into one unbiased member and a smaller string requires *one* augmenting trial if precision is good enough and if the experimental system is assuredly not drifting. The efficiency of parameter estimates is doubled and drift removed if *two* new trials can be done. *Four* augmenting trials—none of them replicates of earlier work—provide still more precision both for the target estimate and for some effects already estimated.

Multiple ambiguities are more likely to require *eight* or more trials. Care in choosing the augmenting set often permits efficient separation of several ambiguities at once, whether or not they were originally in the same string. I always try for sets which permit the P-Q principle (often extended to the P-Q-R-S principle), since this gives the highest efficiency for all estimates. When this cannot be done, because of the asymmetry of the effects found, I patch together the best combination that will permit *some* estimation of the entangled effects. This will, in my experience, nearly always be a new fractional replicate, and not a congeries of unrelated pairs of observations, one pair for each ambiguity.

Although I have no doubt that extensions to the 2^{p-q} for $p > 8$, to the 3^{r-s} series, and to the general augmentation of partially balanced incomplete block systems are all manageable, I leave these pleasures to others.

CHAPTER 15

Trend-Robust Plans

15.1.   INTRODUCTION

This chapter is largely a rephrasing and condensation of a paper published in 1966.* Since no printed evidence of the usefulness of these plans has appeared in nine years, my earlier enthusism for them has been somewhat dampened.

Just as knowledge of a system's stabilities and of its heterogeneities is essential for the effective blocking of a factorial experiment, so similar knowledge is required for systems which drift or trend or age. If the *shape* of the drift curve is roughly known, use should be made of this knowledge. Instead of randomly spreading the systematic trend to give a large random error, we can get most of the trend into two d.f., separated from the random error. This is proved for systems with linear and quadratic trend when a 2^{p-q} plan is to be done, and must, it seems to me, be true also for higher-order trends and for more general factorials.

As Hald [1952b, page 510] has put it: "The possibility of eliminating systematic variation effectively depends on whether the variation is smooth [or whether] . . . irregular fluctuations occur." Later on the same page he recognizes that the trend can be largely eliminated "if the systematic variation

---

* That paper [Daniel and Wilcoxon 1966] fails to point out its obvious indebtedness to two papers by D. R. Cox [1951, 1952] and to Section 14.2 of his book *Planning of Experiments* [1958]. I make this acknowledgment now, with apologies. The basic idea of these plans was F. Wilcoxon's. He noticed that the 8 ordered trials 0 1 1 0 1 0 0 1 gave a sequence for one two-level factor that was exactly orthogonal to linear and quadratic time trends. He then worked out many larger ordered multifactor plans with the same orthogonality.

251

is smooth and slow." I take "smooth" to mean "representable by linear and quadratic terms in time," and "slow" to mean "being a considerable multiple of the time required to make one trial. In the next sections, this multiple goes from 3 to 32.

Some orders of trials for the 2^p, and for the 2^{p-q} are much more sensitive than others to aliasing of linear (L) and quadratic (Q) trends with factorial effect contrasts. Two contrasts in particular (C and BC in the 2^3, D and CD in the 2^4, E and DE in the 2^5) are highly correlated with L and Q, respectively. For B, C, D, E, the squared correlation coefficients with L are 0.80, 0.76, 0.75, and 0.75, respectively. For AB, BC, CD, DE, the squared coefficients with Q are 1.0, 0.80, 0.72, and 0.71, respectively. By choosing *other* contrasts to determine the levels of factors, and by watching out for the corresponding forced 2fi contrasts, we can estimate all main effects and all 2fi's with good efficiencies, using the random fluctuations *about* the drift curve as a basis, rather than the uncontrollable variation including drift.

The factor levels in these plans appear in sequences that may be inconvenient for the experimenter. But there is a wide choice of number of changes of level in each plan, varying, for example, in the 2^4 from 5 to 13. I can see no theoretical objection to allocating the factor hardest to change to the letter that has the longest runs at one level.

Since most estimates are made with efficiencies near 1.00, some experimenters may see an advantage in following the specified sequence even when little is known about the shape of the trend curve. If no trend appears, only two d.f. are lost. If large L- or Q-trends appear, a substantial gain in precision of all estimates is guaranteed.

15.2. SIMPLEST TREND-ROBUST PLANS

Suppose that *one* factor at *two* levels is to be studied in a system known to be drifting linearly over periods as long as several trials. The minimum conceivable plan contains *three* trials. Examples include the effect of some diet change on the milk yield of a milch cow, and the effect of feedstock change on the selectivity of a catalyst that ages slowly in use. Call the two levels of the factor A and B [and not, because of the clumsiness of the designation, (1) and a]. The six possible orders for four trials are:

I.	ABA	IV.	BAA
II.	BAB	V.	ABB
III.	AAB	VI.	BBA

The two orders I and II have obvious advantage. Direct solution of the equation $Y = b_0 + b_1 x + d_1 t$, putting x at -1 and $+1$, and t at -1 and

$+1$, will show that I and II have efficiency 1.00 whereas the other four sequences have efficiencies $\frac{1}{2}$ for b_1 and d_1 and $\frac{2}{3}$ for b_0. Since more than one cow or catalyst batch would have to be used, I and II would naturally be used in alternation.

A severe randomizer will insist that, for each pair, I or II should be chosen at random, to be followed by its complement. Let him.

This miniplan is entirely vulnerable to *quadratic* trend. The only *four*-trial, *one*-factor, *two*-level, *L*- and *Q-free* plan is *ABAB*, which need not be distinguished from its complement, *BABA*. Our fitting equation,

$$(15.1) \qquad\qquad Y = b_0 + b_1 x + d_1 t + d_{11} t^2,$$

requires four constants, so the fitting of four trials (at equal time intervals, please) leaves no room for error estimation. Direct substitution of the four lines of our (X, Y) matrix:

Trial	Spec.	b_0	x	t	t^2	y
1	A	1	-1	-3	9	y_1
2	B	1	1	-1	1	y_2
3	A	1	-1	1	1	y_3
4	B	1	1	3	9	y_4

gives us the following estimates of the four constants in (15.1):

$$
\begin{aligned}
4b_0 &= y_1 + y_2 + y_3 + y_4, & \mathrm{Var}\,(b_0) &= \tfrac{1}{4}; \\
8b_1 &= -y_1 + 3y_2 - 3y_3 + y_4, & \mathrm{Var}\,(b_1) &= \tfrac{5}{16}; \\
8d_1 &= -y_1 - y_2 + y_3 + y_4, & \mathrm{Var}\,(d_1) &= \tfrac{1}{16}; \\
4d_{11} &= y_1 - y_2 - y_3 + y_4, & \mathrm{Var}\,(d_{11}) &= \tfrac{1}{4}.
\end{aligned}
$$

Since the minimum possible variances of the four constants are, respectively, $\frac{1}{4}, \frac{1}{4}, \frac{1}{20}$, and $\frac{1}{4}$, we find efficiency factors of 1, $\frac{4}{5}, \frac{4}{5}$, and 1. Thus, when the *L*- and *Q*-trends account for as little as 25% of the uncontrolled variability, we will gain a more precise estimate of the effect of *A* by using this plan instead of any other.

These two little plans are of course not for actual use except in very rare, probably pedagogical, situations. The efficiency factors for the larger, more realistic plans described below are all greater than $\frac{4}{5}$, with one exception in each plan.

15.3. TREND-ROBUST $2^{2+1}//8$ AND $2^{4-1}//8$

We proceed to give plans and to show their properties, skipping derivations and proofs. Table 15.1 (adapted from Daniel and Wilcoxon [1966]) shows

TABLE 15.1.
A $2^{2+1}//8$, $L-$ AND Q-ROBUST

$$Y = b_0 + b_F x_F + b_G x_G + b_{FG} x_F x_G + d_1 t + d_{11}(t^2 - 5.25)$$

Trial	Spec.			(X)		8×6				$B = (X'X)^{-1}X'$		8×6	
		x_0	x_F	x_G	x_{FG}	t	$(t^2-5.25)$	$8b_0$	$32b_F$	$80b_G$	$16b_{FG}$	$16d_1$	$80d_{11}$
1	(1)	1	-1	-1	1	-3.5	7	1	-3	-7	1	-1	3
2	fg	1	1	1	1	-2.5	1	1	5	11	1	-1	1
3	g	1	-1	1	-1	-1.5	-3	1	-3	9	-3	-1	-1
4	f	1	1	-1	-1	-0.5	-5	1	5	-13	-3	-1	-3
5	(1)	1	-1	-1	1	0.5	-5	1	-5	-13	3	1	-3
6	fg	1	1	1	1	1.5	-3	1	3	9	3	1	-1
7	g	1	-1	1	-1	2.5	1	1	-5	11	-1	1	1
8	f	1	1	-1	-1	3.5	7	1	3	-7	-1	1	3

(X'X)				6×6			$640(X'X)^{-1}$		5×5		
8	0	0	0	0	0						
0	8	0	0	4	0		85	0	-10	-10	0
0	0	8	0	0	-8		0	84	0	0	4
0	0	0	8	-8	0		-10	0	100	20	0
0	4	0	-8	42	0		-10	0	20	20	0
0	0	-8	0	0	168		0	4	0	0	4

Efficiencies: $\frac{16}{17}$ $\frac{20}{21}$ $\frac{4}{5}$ $\frac{16}{21}$ $\frac{20}{21}$

$\hat{F}$ $\hat{G}$ $\hat{FG}$ $\hat{L}$ $\hat{Q}$

$\equiv b_F$ b_G b_{FG} d_1 d_{11}

a $2^{2+1}//8$, along with the ordinary least-squares results, $(X), (X'X), (X'X)^{-1}$, $B = (X'X)^{-1}X'$, and the efficiency factors for B.

The contrasts for $\hat{F}$, $\hat{G}$, $\hat{FG}$, and d_{11} in Table 15.1 support a conjecture made long ago that contrasts with varying but integral coefficients are clarified by being decomposed into two or more contrasts each with the same coefficient, only signs being varied. Thus we have

$$\begin{aligned}
b_F &= -3, 5, -3, 5, -5, 3, -5, 3 \\
&= [-4, 4, -4, 4, -4, 4, -4, 4] \\
&\quad \oplus [1, 1, 1, 1, -1, -1, -1, -1] \\
&= 4(A_0) - (C_0) \\
&= 32\hat{F} - \hat{L}.
\end{aligned}$$

The decomposition here could be found by any idler, but more complex cases can be solved by putting the ordered coefficients themselves through the corresponding forward Yates algorithm. The regression coefficient b_G, to take a more difficult example, looks from its signs like a disturbed $-(AB_0)$. Even if one guessed that $-10(AB_0)$ is the right amount (it is the average of the absolute values of the eight coefficients) and subtracted this to get

$$(X) = -7, 11, 9, -13, -13, 9, 11, -7$$
$$\oplus [10, -10, -10, 10, 10, -10, -10, 10]$$
$$= 3, 1, -1, -3, -3, -1, 1, 3,$$

it still might not be obvious that (X) can itself be partitioned into $2(AC_0) + (BC_0)$. But it is more insightful to note that (X) as it stands is simply $80d_{11}$. Thus b_G is seen to be just the expected factorial contrast [note that g is at its two levels in the 2^{2+1} in the $-(AB_0)$ pattern], corrected for *one* unit of Q ($=d_{11}$) to allow for the small correlation of G with Q.

Responding to natural greed we try to accommodate two more factors into an eight-trial plan, since we have two d.f. left over after fitting F, G, FG, L, and Q. We require firm assurance that the two new factors, H and J, have entirely additive effects. There must be no interaction with H or J. Table 15.2 gives the same results as were shown in 15.1. We have lost a little efficiency in G, but we have gained efficient estimates of the main effects of H and J.

TABLE 15.2.

THE $2^4//8$, $-L$ AND $-Q$

$$Y = b_0 + b_F x_F + b_G x_G + b_{FG} x_F x_G + b_H x_H + b_J x_J + d_1 t + d_{11}(t^2 - 5.25)$$

Trial	Spec.	x_0	x_F	x_G	x_{FG}	x_H	x_J	t	t^2	$\hat{F}$	$\hat{G}$	$\widehat{FG}$	$\hat{H}$	$\hat{J}$	$\hat{L}$	$\hat{Q}$
					(X)	8×8						$32(X'X)^{-1}X' = 32\mathbf{B}$			8×7	
1	(1)	1	-1	-1	1	-1	-1	-3.5	7	-3	-3	2	-4	-2	-2	1
2	fghj	1	1	1	1	1	1	-2.5	1	5	5	2	4	6	-2	1
3	gh	1	-1	1	-1	1	-1	-1.5	-3	-3	3	-6	4	-6	-2	-1
4	fj	1	1	-1	-1	-1	1	-0.5	-5	5	-5	-6	-4	2	-2	-1
5	hj	1	-1	-1	1	1	1	0.5	-5	-5	-5	6	4	2	2	-1
6	fg	1	1	1	1	-1	-1	1.5	-3	3	3	6	-4	-6	2	-1
7	gj	1	-1	1	-1	-1	1	2.5	1	-5	5	-2	-4	6	2	1
8	fh	1	1	-1	-1	1	-1	3.5	7	3	-3	-2	4	-2	2	1

					$(X'X)$	7×7				$128(X'X)^{-1}$			7×7		
8	0	0	0	0	4	0		17	0	-2	0	0	-2	0
0	8	0	0	0	0	-8		0	17	0	0	-2	0	1
0	0	8	0	0	-8	0		-2	0	20	0	0	4	0
0	0	0	8	0	0	0		0	0	0	16	0	0	0
0	0	0	0	8	0	16		0	-2	0	0	20	0	-2
4	0	-8	0	0	42	0		-2	0	4	0	0	4	0
0	-8	0	0	16	0	168		0	1	0	0	-2	0	1

Efficiencies: $\frac{16}{17}$ $\frac{16}{17}$ $\frac{4}{5}$ 1 $\frac{16}{17}$ $\frac{16}{21}$ $\frac{16}{21}$

                $\hat{F}$ $\hat{G}$ $\widehat{FG}$ $\hat{H}$ $\hat{J}$ $\hat{L}$ $\hat{Q}$

15.4. TREND-ROBUST $2^4//16$, $2^{6-2}//16$, $2^5//32$, AND $2^{14-9}//32$

Since the arithmetic for the 16- and 32-run plans will usually be done by computer, we omit the details given in preceding sections and present only the generators of each design:

	$2^4//16$	$2^{6-2}//16$	$2^5//32$	$2^{14-9}//32$
1.	abd	abde	abe	abdegjlm
2.	acd	acdf	ace	acdfgkln
3.	bcd	bcd	bcde	bcdhjklo
4.	abcd	abcdef	d	efgjkl
5.			abcde	abcdefghjklmno

Orderly multiplication of these generators: 1. × 2., 1. × 3., 1. × 2. × 3., 4., etc., for the 2^4 gives (1), *abd, acd, bc, bcd, ac, ab, d, abcd, c, b, ad, a, bd, cd, abc*. Details on confounding patterns and other matters are given in Daniel and Wilcoxon [1966].

CHAPTER 16

Nested Designs

16.1. INTRODUCTION

Most industrial experiments are not done in fully randomized sets, in randomized blocks, or even in randomized incomplete blocks. Even most experiments done after statistical advice do not conform. The reason is one of convenience: some factors are very hard to vary, and others are much easier. It does not make practical sense to vary at random, say, the internal setup of a system operated at high vacuum which must be evacuated and degassed after each takedown. If there are conditions—for example, cathode voltages and emitter temperatures—to be varied within one setup, they will surely be varied while the electrode configuration is maintained constant, without disassembly. *Then* a new assembly will be made, and the easy-to-vary factors again varied inside the newly evacuated system.

It may be advisable to return to the first assembly at some point, but it is almost never technically sensible to test each assembly at only one set of operating conditions. Although no statistical designer would forbid such a plan, he would make sure that the analysis of the resulting data corresponded to the realities of the design.

The fundamental difference between such a *nested* set of data and a fully randomized set lies in their error structures. In the fully randomized plan there is only one homogeneous random error system. Only one error term appears in the modeling equation. No restrictive assumption is made about the sources of the uncontrolled variability. It may come from many physically distinct causes, but it has an equal chance of perturbing each observation. In the nested case, on the other hand, there are at least two independent sources of random disturbance. One set, the *nesting* set, only hits each *group* of data taken together, *once*. In the example used above each system setup is affected by one manifestation of the "setup error." The other set of random causes affects each observation separately as the *nested* factor levels are changed. These *within*-setup perturbations are assumed to be independent of the *among*-setup disturbances. In the simplest nested situations, then, there are two independent sources of random variation. More complicated situations are common.

16.2. THE SIMPLEST CASE

A stratified sample provides the simplest possible example. The data shown in Table 16.1 are taken from Brownlee [1965, Section 10.6, page 325].

TABLE 16.1
NESTED MEASUREMENTS ON 22 BATCHES OF A PLASTIC-LIKE MATERIAL

BROWNLEE'S Table 10.4:

Batch:	1	2	3	4	5	6	7	8	9	10	11
	58	49	45	28	54	47	45	49	43	37	48
	48	41	44	55	49	45	54	47	48	43	52
	47	46	44	50	53	47	50	46	49	47	57
	65	46	44	41	52	47	57	50	47	27	51
Total:	218	182	177	174	208	186	206	192	187	154	208
w:	18	8	1	27	5	2	12	4	6	20	9

Batch:	12	13	14	15	16	17	18	19	20	21	22
	45	55	42	45	41	43	53	41	43	34	50
	43	42	41	43	46	42	44	43	45	34	48
	44	47	46	48	41	38	49	41	44	40	48
	44	52	50	45	30	35	52	35	46	40	48
Total:	176	196	179	181	158	158	198	160	178	148	194
w:	2	13	9	5	16	8	9	8	3	6	2

Here 22 batches of a "plastic-like material" are sampled, each four times. A trial model equation may be written as

(16.1) $y_{i(j)} = \mu + e_i + \varepsilon_{i(j)}$, $i = 1, \ldots, 22; (j) = 1, \ldots, 4$ for each i.

If the sampling of both batches and subbatches is random, then

$$E\{e_i\} = E\{\varepsilon_{i(j)}\} = E\{e_i e_{i'}\} = E\{\varepsilon_{i(j)}\varepsilon_{i(j')}\} = E\{e_i\varepsilon_{i(j)}\} = 0.$$

The random errors are all assumed to be unbiased and uncorrelated. The usual criticism of experimenters—that they do not study their system under a sufficiently wide range of experimental conditions—is here replaced by the usual criticism of samplers—that they overdo the estimation of the nested random error at the expense of the nesting error. There are to be $3 \times 22 = 66$ d.f. for estimating Var $(\varepsilon_{i(j)}) = E\{\varepsilon_{i(j)}^2\} = \sigma_0^2$ in this case, and less than 21 for estimating Var $(e_i) = E\{e_i^2\} = \sigma_1^2$.

We will get an estimate of σ_0^2 from

$$\frac{1}{66} \sum_{i=1}^{22} \sum_{j=1}^{4} (y_{i(j)} - \bar{y}_{i.})^2 = A,$$

that is, from the pooled sum of squares *within* batches. But the mean square (among batches),

$$\frac{1}{21} \sum_{i=1}^{22} (\bar{y}_{i.} - \bar{y}_{..})^2 = B,$$

has the expected value $(\sigma_1^2 + \sigma_0^2/4)$ and so must be used with A to get an estimate of σ_1^2.

We have plunged ahead and made statements about the standard way to estimate σ_0^2 and σ_1^2 in nested samples, with no warnings or cautions. When data are taken in this standard way, the same number of subbatches being inspected from each batch, it is not difficult to examine the so-far-unspoken assumptions of normality and of homogeneity of the two sets of random effects. For the $\varepsilon_{i(j)}$ the *ranges* of the y_{ij} over i should follow the known distribution of ranges of four from a normal population. Figure 16.1 shows by its solid curve the expected distribution of normal ranges of four, with $\hat{\sigma}_0$ taken as $4.2 =$ average range/$d_2 = 8.68/2.06$, where d_2 is taken from Hald's Table VIII [1952a, page 60]. This seems to me a clear case of a smooth nonnormal distribution, not due to one or a few outlying observations. A smooth curve can be drawn through the observed cumulative distribution of the 22 ranges, but it crosses the expected curve only once, all

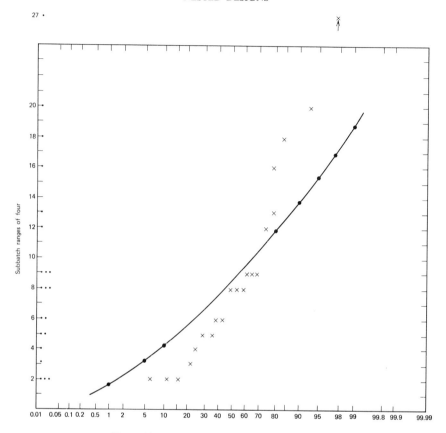

Figure 16.1   X, E.c.d. of 22 ranges of sets of four.

the larger ranges being too large, all the smaller ones too small. To see whether there is any clear dependence of range (and hence of σ_0 or σ_0^2) on batch mean, refer to Figure 16.2a. I see no connection whatever. In desperation I try transforming the observations $y_{i(j)}$ by taking logarithms, square roots, and reciprocals, and by deleting the four largest observations. None of these rectifies the nonnormality visible in the figure, and so none of them is reproduced. The reason for trying to find a (simple) transformation to normality is that we want to make a test of significance and even a confidence-interval statement about the two components of variance. But these are derived under the assumptions that the e_i and the $\varepsilon_{i(j)}$ are normally and independently distributed. We try doubly hard here because, as Figure 16.3 shows, the batch means (totals) are nearly normal.

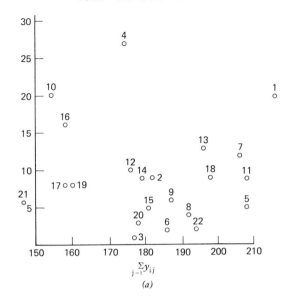

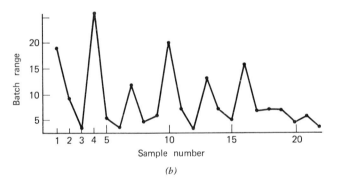

Figure 16.2 (*a*) Twenty-two batch ranges (W_i) versus batch totals ($\sum y_j)_i$. See Table 16.1.

On plotting the 22 ranges of four in serial order (Figure 16.2*b*), a glimmer emerges. Every third range is higher than its two successors. I guess therefore, that we are seeing the effects of successive subbatch results from some *mixing* process. The batch mean is not changing, but the within-batch homogeneity is changing rapidly in cycles of three batches. I have divided the (first 21) batches into three sets of seven, found the average range of each set (namely, 16.29, 6.57, 4.43), deduced a standard deviation by dividing by 2.06 (to get 7.91, 3.19, 2.15), standardized each set by dividing by its corresponding

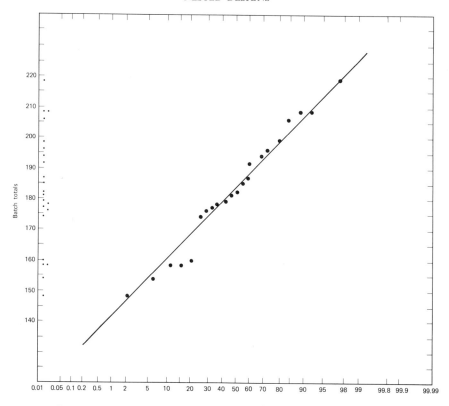

Figure 16.3   E.c.d. of twenty-two batch totals. $s_B = (200 - 165)/2 = 17.5$.

standard deviation, and plotted the whole set of 21 "tristandardized ranges" in Figure 16.4. As can be seen, we have a success of sorts.

Although I do not like the ripples in the e.c.d., they are not wide and the extreme ranges now behave well compared to their mates. A more careful and more sensitive test would take the residuals *within* each batch, group them into three sets, and produce three new e.c.d.'s. Let's do it. Table 16.2 gives the grouped residuals, and Figures 16.5a, b, c show the e.c.d. on normal grids for each set of 28 separately. The error within each "mixing stage" is homogeneous, nearly normal, and so acceptable. But the model equation (16.1) must be used separately on each stage since each represents a drastically different relation between the two components of variance, σ_0^2 and σ_1^2.

We now feel safe in partitioning the total sum of squares for each set of seven batches (TSS_k, $k = $ I, II, III), using the usual identity:

$$(y_{ij} - \bar{y}_{..})_k = (y_{ij} - \bar{y}_{i.})_k + (\bar{y}_{i.} - \bar{y}_{..})_k, \qquad k = \text{I, II, III}.$$

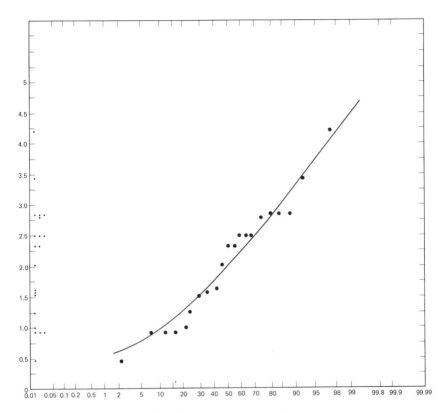

Figure 16.4 Twenty-one tristandardized ranges.

TABLE 16.2

RESIDUALS FROM 21 BATCH MEANS GROUPED INTO THREE SETS [BROWNLEE 1965, PAGE 325]

i	I				i	II				i	III			
1	3.5,	−6.5,	−7.5,	10.5	2	3.5,	−4.5,	0.5,	0.5	3	0.7,	−0.2,	−0.2,	−0.2
4	−15.5,	11.5,	6.5,	−2.5	5	2.0,	−3.0,	1.0,	0	6	0.5,	−1.5,	0.5,	0.5
7	−6.5,	2.5,	−1.5,	5.5	8	1.0,	−1.0,	−2.0,	2.0	9	−3.8,	1.2,	2.2,	0.2
10	−1.5,	4.5,	8.5,	−11.5	11	−4.0,	0 ,	5.0,	−1.0	12	1.0,	−1.0,	0 ,	0
13	6.0,	−7.0,	−2.0,	3.0	14	−2.8,	−3.8,	1.2,	5.2	15	−0.2,	−0.2,	2.8,	−0.2
16	1.5,	6.5,	1.3,	−9.5	17	3.5,	2.5,	−1.5,	−4.5	18	3.5,	−5.5,	−0.5,	2.5
19	1.0,	3.0,	1.0,	−5.0	20	−1.5,	0.5,	−0.5,	1.5	21	−3.0,	−3.0,	3.0,	3.0

263

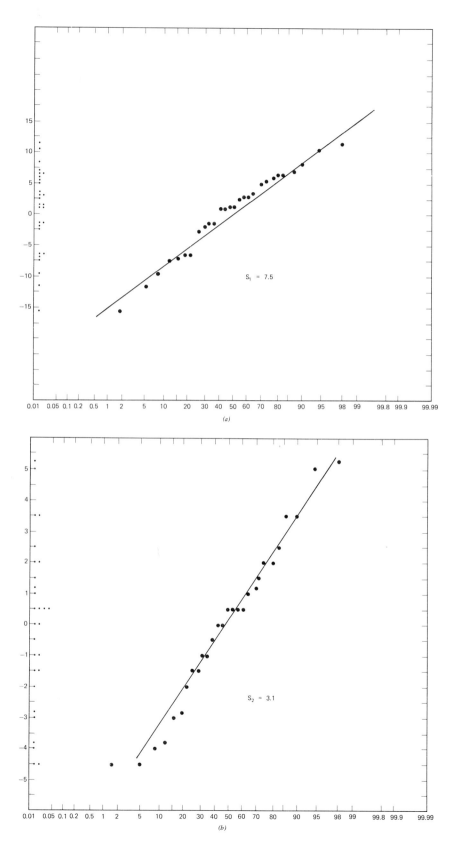

(a)

$S_1 = 7.5$

(b)

$S_2 = 3.1$

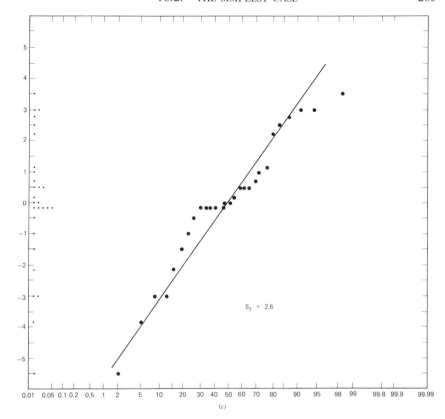

$s_3 = 2.6$

(c)

Figure 16.5 (a) E.c.d. of residuals from seven batches $(1, 4, 7, \ldots, 19)$, $s_1 = 7.5$. (b) Same for seven batches $(2, 5, 8, \ldots, 20)$, $s_2 = 3.1$. (c) Same for seven batches $(3, 6, 9, \ldots, 21)$, $s_3 = 3.6$.

Also,

$$\text{TSS}_k = \sum_{i=1}^{7} \sum_{j=1}^{4} (y_{ij} - \overline{y}_{i.})^2 = \sum\sum(y_{ij} - \overline{y}_{i.})^2 + 4 \sum_{i=1}^{7} (\overline{y}_{i.} - \overline{y}_{..})^2$$
$$= \text{SS (within batches)}_k + 4 \text{ SS (among batches)}_k$$
$$= \text{SS } W_k + 4 \text{ SS } B_k.$$

Since $E\{\text{MS }(W_k)\} = \sigma_{0,k}^2 + 4\sigma_{1,k}^2$, we compute

$$\text{MS }(W_k) = \frac{\text{SS }(W_k)}{21} \quad \text{and} \quad \text{MS }(A_k) = \frac{\text{SS }(A_k)}{6}.$$

TABLE 16.3.

k	MS (B_k)	MS (W_k)	$\hat{\sigma}^2_{1,N}$	$F = \text{MS}(B)/\text{MS}(W)$	P	$\hat{\text{d.f.}}_B$	L	U
1	165.29	58.14	26.79	2.84	.05	2.4	0.07	2.77
2	79.74	9.50	17.56	8.39	.0005	4.6	0.21	2.27
3	60.50	2.55	14.49	23.73	.0005	5.5	0.25	2.15

Table 16.3 shows the mean squares and variance component estimates. The "equivalent degrees of freedom for $\hat{\sigma}^2_1$," called $\hat{\text{d.f.}}_B$ in the table, is computed by Satterthwaite's formula* [1946], when and only when the F-value reaches Gaylor and Hopper's criterion [1969], namely, when its significance probability is below 0.025. For the present case, where f_1 is 6 and f_2 is 21, this value is 3.09, and only sets II and III ($k = 2, 3$) meet the requirement. The equivalent degrees of freedom are then 4.64 and 5.50—not too far below their maximum possible value of 6.

Once we have $\hat{\text{d.f.}}_B$ *and* once we have some assurance that our random distributions are normal, we can put confidence ranges around $\sigma^2_{B,K}$. As all good elementary textbooks aver (e.g., Brownlee, [1965], Section 9.2, pages 282 ff.; Davies [1971], Section 2.363, pages 28 ff; Hald, [1952b], Section 11.4, page 286), we need only multiply our estimates, $s^2_{B,k}$ by $\hat{\text{d.f.}}/\chi^2_{.95}$ and by $\hat{\text{d.f.}}/\chi^2_{.05}$ to get 90% intervals for $\sigma^2_{B,k}$. These two values for our case are 0.75 and 4.75. These limits serve mainly to warn us that we do not know our among-batch components of variance very well. This can hardly be news at this stage. A variance estimated with fewer than 6 d.f. is of course poorly estimated. Even that estimated by Brownlee [1965, page 327] with something less than 21 d.f. is, as he says, of disagreeably poor precision.

The conclusions for this set of data are now tolerably clear. The 22 sets-of-four measurements were not taken from a system with homogeneous error. When they are divided into three subsets of seven, we find reasonably constant random normal error within each subset but widely different sub-batch variation from set to set. It should not be argued by a critic of this finding that we have deliberately divided the data into sets with visibly different subbatch errors. We have dared to do this only because of the

---

*This formula may be written, for our case:

$$\text{d.f.}_B = \frac{(F - 1)^2}{\dfrac{F^2}{f_1} + \dfrac{1}{f_2}}$$

where F is the usual ratio: $\text{MS}(B)/\text{MS}(W)$,
        f_1 and f_2 are the numerator and denominator d.f.,
        d.f._B is the approximated d.f. for $\hat{\sigma}^2_B$.

completely regular periods of three in the successive batches. We do seem now to be looking at three nested cases, each with its own within-nest (i.e., with subbatch) normal error with variances 58.14, 9.50, and 2.55, rather than at one jumble with average variance 23.39.

It is some consolation, I should hope, that there is more information about the among-batch variation in the 7 batches, 3, 6, 9, 12, 15, 18, 21, than in the other 14—or, of course, than in the whole scrambled set. We have not destroyed the data; we have saved them, and found out which part is worth saving.

The general conclusions are methodological:

1. One has little chance of finding anything objective about the system that produced the data being analyzed if one declines to check the correctness of the crucial assumptions required by the estimation and test procedures being used.
2. There do not seem to be entirely general rules for studying such data. There must be some more efficient ways than my own—I have spent most of 3 weeks looking at these 88 values. Each time that I try to write down a set of general rules, they seem to me to depend too heavily on the last set of data I have studied. Just now I would recommend, for balanced data the following:

    a. Write down in detail what is wanted from the data. Write down how you would attain these desiderata if all is well. Write down the assumptions that *must* be valid if the statistics you would normally compute are to be valid. See how many of these assumptions you can check from these data alone. See how many of those which you cannot check can be guessed, guaranteed, or refuted by the experimenter who secured the data.
    b. Make all simple graphs: data in time order, internal versus external means, ranges, and e.c.d.'s of all subsets with more than 15 d.f., especially if normality is required.

The title of this section is misleading. I have left it unchanged since it more nearly reflects life as one thinks it is: this was an *apparently* simple case.

16.3. SPLIT-PLOT DESIGNS

We have just discussed a case of simple nesting. Both systems of random error—that within batches and that among batches—affected each observation, but aside from the estimation of the two average (squared) random disturbances only *one* parameter of the system, its grand average, was estimated. No factors were varied, so this was not really an experiment.

We come now to the much commoner case, in which *some* factor or factors will be varied among batches (or plots), while some *other* factor or factors will be varied *within* each batch or plot. There is an excellent discussion of such an experiment in Cochran and Cox [1957, pages 293–302] with a detailed example. In studying chocolate cakes, *three* recipes were used, at *six* baking temperatures, for each of *15* replicate "mixes" or batches. The baking temperature was varied (on separate subbatches) within each batch. Thus temperature effects and all their interactions were affected only by subplot random variation, whereas recipes had the whole-plot error. The recipe × replication interaction was used to estimate the whole-plot error.

The detailed analysis of the sort indicated above for simple nesting is left as an exercise for the reader, who will notice that some 28% of the SS for whole-plot error is produced by *two* cakes (Nos. 4 and 14 in Recipe I) and that, if these two are revised, the whole-plot error MS is reduced nearly to insignificance. There is again, then, no possibility of estimating the precision with which the whole-plot variance is estimated, since even the F-ratio of 1.63 for the unretouched whole-plot error is too small for Satterthwaite's formula to be valid.

There are also good examples in Cox [1958], Johnson and Leone [1964, Vol. II, page 232], Federer [1955, page 276], and Hicks [1965, pages 192 ff.]. Two of these are briefly discussed here.

The data given by D. R. Cox are described as a *possible* set (my emphasis) and are used only to show the rationale and mode of interpretation for plans in split *units*, as Cox calls them. His discussion of this experiment and of other split-unit designs is well worth studying. There are irregularities in the data (Day 2 does not seem like the other 7 days; there is an upsetting three-factor interaction) that dash once again my hopes for a designed experiment that produces both precise and simple results, but there is no use crying over spilt plots.

The split-plot example in C. R. Hicks's book is accompanied by a clear and general discussion, including a straightforward method of deducing the expected values of all mean squares and hence a means of seeing which effects are testable by each error type. The data are given in Table 16.4*a*. The plots are the sets of three pieces at one temperature within one replicate. They were electronic components withdrawn from one oven at three bake times. Thus all interactions with bake time, as well as the main effect, are measured with within-plot precision. The *among*-plot MS for error (296) appears to be indistinguishable from the within-plot MS (243), and therefore the main technical gain must have been the ease of putting three components at a time into one oven.

TABLE 16.4a, b

a. HICKS'S TABLE 13.3: SPLIT-PLOT LAYOUT FOR ELECTRICAL COMPONENT-LIFE TEST DATA

Rep. (R)	Baking Time (B), min	Oven Temperature (T), degrees				Averages
		580	600	620	640	
I	5	217	158	229	223	207
	10	233	138	186	227	196
	15	175	152	155	156	158
II	5	188	126	160	201	169
	10	201	130	170	181	170
	15	195	147	161	172	109
III	5	162	122	167	182	158
	10	170	185	181	201	184
	15	213	180	182	199	194

b. HICK'S TABLE 13.5: ANOVA* FOR SPLIT-PLOT ELECTRICAL COMPONENT DATA

Source	d.f.	SS	MS	EMS
R_i	2	1,963	982	$\sigma_e^2 + 12\sigma_R^2$
T_j	3	12,494	4165	$\sigma_e^2 + 3\sigma_{RT}^2 + 9\sigma_T^2$
RT_{ij}	6	1,774	296	$\sigma_e^2 + 3\sigma_{RT}^2$
B_k	2	566	283	$\sigma_e^2 + 4\sigma_{RB}^2 + 12\sigma_B^2$
RB_{ik}	4	7,021	1755	$\sigma_e^2 + 4\sigma_{RB}^2$
TB_{jk}	6	2,601	434	$\sigma_e^2 + \sigma_{RTB}^2 + 3\sigma_{TB}^2$
RTB_{ijk}	12	2,912	243	$\sigma_e^2 + \sigma_{RTB}^2$
Total:	35	29,331		

* Analysis of variance.

Although the author writes that the "results of the analysis" are shown in the analysis of variance table (Table 16.4b), my own emphasis is that the table tells us only which data subtables are worth examining. The strikingly large $R \times B_{ik}$ mean square tells me to look at the $R \times B$ table of averages or totals. This is given at the right of Table 16.4a. It tells something rather unsettling. The effect of increasing bake time was *negative* and roughly linear in Replicate I; it was zero in Replicate II; it was *positive* and roughly linear in Replicate III. This is not the way in which replicates are suppose to behave.

The reader may feel that my attitude is antiscientific, or even perverse, in that I am disturbed when experimental results do not come out consistently. But we do replicates to gain precision; if they show different patterns, and if we take the data seriously, then we can only point out to the experimenter that just when his error is least (in Replicate II) his bake-time effects vanish, and that he may want to look into his technique and records quite carefully.

16.4.   FRACTIONAL FACTORIALS, 2^{p-q}, IN SPLIT-PLOT ARRANGEMENTS

The simplest nested two-level plan is the 2^2. Imagine, then, that only two two-level factors are to be studied, *and* that A is hard to vary whereas B is easy. We must now think of doing a 2^2 in two blocks of two:

I	II
(1)	a
b	ab

We can estimate the average B-effect by the usual contrast, and it is clear, is it not, that the *difference* between the two within-block B-effects measures four times the AB interaction with the same, that is, with σ_0^2, the within-block, variance. The A-effect is measured by a contrast across blocks and hence has variance $\sigma_1^2 + \sigma_0^2/2$. Replicates must be done, of course, to get estimates of the two variances, as well as to gain more precision in estimating all effects.

We can view each block or plot as a 2^{2-1}, with alias subgroup $I \pm A$. We may describe a split-plot plan as one in which some main effect(s), here only A, is (are) aliased with block means.

As more factors are considered, more alternatives for plot splitting become available. For the 2^3 we might have blocks of four, with two factors varied inside plots, and so have each block a 2^{3-1} with, say, A again aliased with blocks. Here too the alias subgroup for plot I is $I - A$ and, for plot II, $I + A$:

I	II
(1)	a
b	ab
c	ac
bc	abc

It is obvious that B, C, and BC are estimated within plots. It is not quite so obvious that AB, AC, and ABC are also estimated with within-plot precision. For example, AB, which is always

$$(1) - a - b + ab + c - ac - bc + abc,$$

can be calculated from

$$(abc + ab - a - ac) - [bc + b - (1) - c]$$
$$= B\text{-effect at high } A - B\text{-effect at low } A,$$

that is, from within-plot contrasts, and hence with the within-plot variance. Only A is measured with among-plot, that is, with whole-plot, variance.

A second alternative for the 2^3 might demand blocks of two with C for the within-plot factor and with A and B varied among plots. The single plot, I, is now a 2^{3-2} with alias subgroup

$$I - A - B + AB,$$

whence the aliased effects within plots are

$$C - AC - BC + ABC = E\{c - (1)\}.$$

The four plots are as follows:

I	II	III	IV
(1)	a	b	ab
c	ac	bc	abc

Here it should be clear or be deducible by the reader that A, B, and AB are estimated with the "outer" variance; the other effects, with the inner variance. And again it is taken for granted that the whole 2^3 will be run more than once to get estimates of the two variance components.

Consider now a 2^5 with three factors nested. We should perhaps write this as a $2^2 \times 2^{(3)}$ to separate the whole-plot factors from the others, and to make our example one with plots containing four treatment combinations. The principal plot, if we may call it that (and we may), is I below:

I	II	III	IV
(1)	a	b	ab
cd	acd	bcd	$abcd$
ce	ace	bce	$abce$
de	ade	bde	$abde$

Since we have outlined only half of a full 2^5, we cannot hope to separate all 2fi's. The reader can see that CDE is aliased with the mean of all four plots, so that the overall a.s.g. is $I - CDE$. Thus the three 2fi's, CD, CE, and DE, are not separable from their complementary main effects. This half replicate would, then, be done only if C, D, and E were known to have additive effects. We require the other 2^{5-1}, $I + CDE$, containing all treatments with *odd* numbers of the letters c, d, e, if we are to separate all interactions. We will still have A, B, and AB estimated with whole-plot error.

It must have occurred to the reader that the greater convenience of plot splitting entails a price that may be excessive. Two variances must be estimated, and the number of degrees of freedom available for the two estimates can be made equal only when plots are of size 2. As the reader can see from the many text examples, the number of degrees of freedom usually increases rapidly as one goes down to the inner components. This may be good for the subplot effects but is of course correspondingly bad for the whole-plot effects.

All of the above discussion may be taken as an introduction to the useful paper of Addelman [1964], which is reprinted in Appendix 16.A. A few nomenclatural differences should be noted. Addelman uses P, Q, R, ..., W for subplot factors; he uses the symbol $PQRS_{0 \text{ or } 1}$ where I would write $\pm PQRS$, and his $PQR_{0,1}$ corresponds to my $\mp PQR$; finally he calls the alias subgroup the "identity relationship."

Addelman writes [1964, page 255, four lines below Table 2], "If some or all of the 3fi ... are known to be negligible, they may be pooled with the interactions that are used to measure experimental error." This is unexceptionable, but I have yet to meet an experimeter who knew that his 3fi's (or even his 2fi's!) were negligible. I would then advise looking at the results of the experiment, pooling everything that looked as small as the higher interactions. As I have so often said before, most 3fi's (by which I mean far more than 90%) are found to be indistinguishable from random error.

We take as an example the earliest published split-plot 2^{p-q} fractional replicate [Kempthorne and Tischer 1953] (see Tables 16.5a, b), even though it is further complicated in two ways. In the first place, it was actually an $8 \times 4^2 \times 2^4$ in 512 treatment combinations in which the seven d.f. for three "pseudofactors" and all their interactions were used to represent the main effects of one eight-level factor, and two extra pseudofactors were used for each of the four-level factors. The second complication was that the plan actually had its subplots split into sub-subplots, and so, for those whose tongues are agile, it was a split-split-plot design with three error terms.

Although it is not directly relevant to the split-plot aspects of this experiment, the energetic reader will want to verify that half of the 2fi variety × date (the 8×4 table of means is given in Table 16.5b) interaction MS, which is judged highly significant, comes from a single cell (7th row, 1st column) where a 3.63 average appears. If it were 2.63, the 2fi would not be significant!

A very energetic reader may notice that the SS's computed from the 8 × 4 table appear to be less than those in Kempthorne's analysis of variance table by a factor of 36. Mr. Kempthorne has kindly resolved this mystery. Each of the treatment combinations was actually the sum of 6 judgments, two from each of three judges.

TABLE 16.5a, b.
TABLES FROM KEMPTHORNE AND TISCHER'S 8 × 4² × 2⁴ IN 512 TREATMENTS

a. KEMPTHORNE AND TISCHER'S TABLE 2: ANALYSIS OF VARIANCE FOR APPEARANCE

Source of Variation	Degrees of Freedom	Sum of Squares	Mean Square
Rep.	3	38.71	
Var (A, B, C)	7	1190.19	170.03†
Error A	21	635.23	30.25
Total	31	1,864.13	
Date (D, E)	3	9046.14	3015.38†
Date × Var	21	1502.80	71.56†
FG	1	116.28	116.28†
HJ	1	578.00	578.00†
Error B	70	2787.96	39.83
Total	127	15,895.31	
F	1	36.12	36.12
G	1	12.50	12.50
H	1	5189.26	5189.26†
J	1	2601.01	2601.01†
FH	1	0.94	0.94
FJ	1	29.07	29.07
GH	1	23.63	23.63
GJ	1	1.76	1.76
Var × F	7	51.32	7.33
Var × G	7	146.63	20.95
Var × H	7	418.62	59.80*
Var × J	7	722.62	103.23†
Date × F	3	61.59	20.53
Date × G	3	39.02	13.01
Date × H	3	467.82	155.94†
Date × J	3	141.98	47.33
Error C	336	7303.68	21.74
Total	511	33,142.88	

TABLE 16.5 (*continued*)

b. KEMPTHORNE AND TISCHER'S TABLE 3: VARIETY AND DATE EFFECTS AND INTERACTIONS

Variety	Date 1	2	3	4	Mean
1	2.56	3.59	4.45	5.00	3.90
2	3.19	4.23	4.53	5.10	4.26
3	3.51	5.04	5.17	4.92	4.66
4	3.09	4.34	4.68	5.28	4.35
5	2.78	3.92	5.08	5.01	4.20
6	3.24	3.89	4.97	5.17	4.32
7	3.63	3.50	4.05	3.95	3.78
8	3.08	4.06	4.83	5.07	4.26
Mean	3.14	4.07	4.72	4.94	4.22

Standard errors: of variety means: 0.12
of date means: 0.09
of entries in table for interaction: 0.26

Average Effects of Blanching, Dehydration, and Storage

	F Blanching Temperature	G Dehydration Temperature	H	J Storage 3 Months	6 Months	Mean
Low:	4.26	4.24	70°F	5.30	4.19	4.75
High:	4.17	4.19	100°F	3.88	3.49	3.69
				4.59	3.84	

* Significant at .05 level
† Significant at .01 level

16.5. PARTIALLY HIERARCHAL PLANS

I have neglected quite a number of important matters in this work. One major omission has been "mixed models." This term is meant to cover experimental situations in which one factor has fixed levels, and one has random levels, being, say a sample from some population, finite or infinite. When a mixed model situation is investigated by a split-plot design, we have a "partially hierarchal plan." The earliest sizable example of such a "p.h.p."

was reported in two related papers by Vaurio and Daniel [1954] and by Scheffé [1954]. Our detailed model was so greatly improved by Wilk and Kempthrone [1955] that it is now entirely out of date. The example is persistent, however, and has been discussed repeatedly ever since. I do not think that the last word has been said. Perhaps a new controversy is in the making. It would hardly be useful to write an exposition now that may well be obsolete before it reaches print. Those who must do something in the meantime are advised to read Scheffé rather than Brownlee. [The reason for this recommendation becomes clear when one reads in Scheffé that Bennett and Franklin (whom Brownlee follows) use a correct method, but from an incorrect derivation. Scheffé's exposition is difficult, but there we are.]

16.6. SUMMARY AND CONCLUSIONS

Nested designs (some factors held constant, others varied in each "nest") are common in industrial research. The larger the system under study, the more likely it is that such plans will prove the more convenient or even the only possible ones. Their disadvantages are low replication for the factors held constant for each nest, and the consequent loss of degrees of freedom for estimating variances of the effects of the nesting factors.

The obligation to examine the data from nested designs (for defective points and for nonnormal error distribution) is even more binding than for other designs. Since we usually want to estimate each component of variance (by subtraction of one mean square from another), the assumption of normality of residual distribution is a key one. It is clearly violated in some published examples.

APPENDIX 16.A

SOME TWO-LEVEL FACTORIAL PLANS WITH SPLIT PLOT CONFOUNDING*,†

SIDNEY ADDELMAN

Research Triangle Institute

The classical split plot experiment consists of plots within blocks and blocks within replicates. The blocks are usually referred to as "whole plots" and the plots as "split plots." In most split plot experiments, the whole plots

* Reprinted by permission from Addelman [1964].

† This research was supported by the U. S. Army Research Office (Durham).

are split into a number of parts equal to the number of split plot treatments. It will frequently be found in practice that the number of split plot treatments available exceeds the number of split plots. Assume that there are four whole plot treatments and eight split plot treatments, the split plot treatments being the treatment combinations of three two-level factors. It is usually reasonable to assume that the three-factor interaction effect of the three split plot factors is of less interest than the other split plot comparisons. When this is the case, the three-factor interaction may be confounded with whole plots. The confounding of split interaction effects with whole plots is known as split plot confounding. Examples of split plot confounding have been presented by Kempthorne.* If the whole plot treatments are denoted by t_1, t_2, t_3, and t_4, and the split plot factors are denoted by P, Q, and R, one complete replicate of the experimental plan would involve eight whole plots, each split into four split plots, as follows, before randomization:

t_1	t_1	t_2	t_2	t_3	t_3	t_4	t_4
000	001	000	001	000	001	000	001
011	010	011	010	011	010	011	010
101	100	101	100	101	100	101	100
110	111	110	111	110	111	110	111

The treatment combinations 000, 011, etc., denote the combination of levels of factors P, Q, and R, respectively. If we denote those treatment combinations for which the sum of the levels of the factors P, Q, and R is 0 (modulo 2) by PQR_0 and 1 (modulo 2) by PQR_1, the experimental plan can be represented as follows:

t_1	t_1	t_2	t_2	t_3	t_3	t_4	t_4
PQR_0	PQR_1	PQR_0	PQR_1	PQR_0	PQR_1	PQR_0	PQR_1

It is clear that the PQR interaction is confounded with the whole plot treatment replicates. The analysis of variance for this plan will have the structure shown in Table 16A.1.

---

* O. Kempthorne, Recent developments in the design of field experiments, *Journal of Agricultural Science*, 37: 156–162 (1947); *The Design and Analysis of Experiments*, John Wiley & Sons, Inc., New York (1952).

TABLE 16A.1
STRUCTURE OF ANOV FOR
A 2^5 FACTORIAL PLAN WITH
SPLIT PLOT CONFOUNDING

Source	d.f.
Whole plot treatments: T	3
PQR	1
$TPQR$	3
Split plot treatments: $P, Q, R,$ PQ, PR, QR	6
TP, TQ, TR TPQ, TPR, TQR	18
Total	31

If the higher order interactions are negligible, the $TPQR$ interaction can be used as an estimate of whole plot error with three degrees of freedom and the $TPQ, TPR,$ and TQR interactions can be pooled to form an estimate of split plot error with nine degrees of freedom.

There are many situations in which the whole plot treatments are the combinations of two-level factors. For example, the four whole plot treatments in the experimental plan already described might be the four treatment combinations 00, 01, 10, and 11 of the whole plot factors A and B. In such a situation, the three degrees of freedom for whole plot treatments, T, can be partitioned into single degree of freedom contrasts, denoted by $A, B,$ and AB.

It should be noted that the experimental plan consists of a full replicate of a 2^5 factorial arrangement, two of the factors representing whole plot treatments and the remaining three factors representing split plot treatments. Split plot confounding can also lead to fractional replicates of the factorial arrangements. Consider, for example, a situation in which two factors, each at two levels, are to be tested on whole plots while the treatment combinations of a 2^5 factorial arrangement are to be tested on split plots. If the whole plot factors are denoted by A and B while the split plot factors are denoted by $P, Q, R, S,$ and T, an experimental plan with eight whole plots, each split into eight split plots, could be represented as follows before randomization:

00	00	01	01	10	10	11	11
$PQRS_0$	$PQRS_0$	$PQRS_1$	$PQRS_1$	$PQRS_1$	$PQRS_1$	$PQRS_0$	$PQRS_0$
PQT_0	PQT_1	PQT_0	PQT_1	PQT_0	PQT_1	PQT_0	PQT_1
RST_0	RST_1	RST_1	RST_0	RST_1	RST_0	RST_0	RST_1

The interaction $PQRS$ is confounded with the AB whole plot contrast, while PQT and RST are confounded with whole plot replicates. The 64 treatment combinations of the plan constitute a half replicate of a 2^7 arrangement (a 2^{7-1} plan) defined by the identity relationship $I = ABPQRS_0$. A systematic procedure for obtaining the treatment combinations that are defined by an identity relationship was presented by Addelman.*

The analysis of variance for the above plan will have the structure shown in Table 16A.2.

TABLE 16A.2
STRUCTURE OF ANOV OF A 2^{7-1} FACTORIAL PLAN
WITH SPLIT PLOT CONFOUNDING

Source	d.f.
$A, B, AB + PQRS$	3
PQT, RST	2
Whole plot error: $APQT + BRST, ARST + BPQT$	2
$P, Q, R, S, T, PQ, PR, PS, PT,$ QR, QS, QT, RS, RT, ST	15
$AP, AQ, AR, AS, AT, BP, BQ, BR, BS, BT$	10
$PRT, PST, QRT, QST, ABT, APT, AQT,$ $ART, AST, BPT, BQT, BRT, BST$	13
$ABS + PQR, ABR + PQS, ABQ + PRS, ABP + QRS,$ $APQ + BRS, APR + BQS, APS + BQR, ARS + BPQ,$ $AQS + BPR, AQR + BPS$	10
Split plot error: $ABPT + QRST, ABQT + PRST,$ $ABRT + PQST, ABST + PQRT, APRT + BQST,$ $AQST + BPRT, APST + BQRT, AQRT + BPST$	8
Total	63

In this table, all five-factor and higher order interactions have been ignored. The whole plot and split plot error terms are made up of sets of aliased four-factor interactions, because it is likely that these four-factor interactions are negligible. If some or all of the three-factor interactions or pairs of three-factor interactions are known to be negligible, they may be pooled with the interactions that are used as an estimate of experimental error.

A partial index of two-level factorial and fractional factorial arrangements that involve split plot confounding is presented in Table 16A.3. The whole plot treatments are the treatment combinations of n_1 (= 2, 3, 4) two-level factors, and the split plot treatments are the combinations of levels of n_2 (= 3, 4, . . . ,

* S. Addelman, Techniques for constructing fractional replicate plans, *Journal of the American Statistical Association*, 58: 45–71 (1963).

TABLE 16A.3
Partial Index of 2^n Factorial Plans Involving Split Plot Confounding

Plan No.	No. of Whole Plots	No. of Split Plots	Whole Plot Factors	Split Plot Factors	Identity Relationship Generators	Confounding* Generator
1	4	4	A, B	P, Q, R	ABPQR	
2	4	8	A, B	P, Q, R, S	ABPQRS	
3	4	16	A, B	P, Q, R, S, T	ABPQRST	
4	4	16	A, B	P, Q, R, S, T, U	APQRS, BPQTU	
5a	4	32	A, B	P, Q, R, S, T, U, V	APQRST, BPQRUV	
5b	4	32	A, B	P, Q, R, S, T, U, V	PQRSTUV, ABPQR	
6	4	32	A, B	P, Q, R, S, T, U, V, W	PQRSTUVW, APQRS, BPQTU	PQR
7	8	4	A, B	P, Q, R, S	ABPQRS	PQRS
8	8	8	A, B	P, Q, R, S, T		PQT
9	8	8	A, B	P, Q, R, S, T	APQRS, BPQTU	QRT
10	8	8	A, B	P, Q, R, S, T, U	APQRST, BPQRUV	QSU
11	8	16	A, B	P, Q, R, S, T, U, V	PQRSTUVW, APQRS, BPQTU	PRUV
12	8	16	A, B	P, Q, R, S, T, U, V, W	ABCPQR	
13	8	4	A, B, C	P, Q, R	ABCPQRS	
14	8	8	A, B, C	P, Q, R, S	ABPQR, ACPST	
15	8	8	A, B, C	P, Q, R, S, T	ABPQRS, ACPQTU	
16	8	16	A, B, C	P, Q, R, S, T, U	ABPQR, ACPST, ABCUV	
17a	8	16	A, B, C	P, Q, R, S, T, U, V	PQRSTUV, ABPQR, ACRST	
17b	8	16	A, B, C	P, Q, R, S, T, U, V	PQRSTUVW, APQRS, BPQTU, CPRUV	
18	8	16	A, B, C	P, Q, R, S, T, U, V, W	ABCD, ABPQR	
19	8	4	A, B, C, D	P, Q, R	ABCD, ABPQRS	
20	8	8	A, B, C, D	P, Q, R, S	ABCD, ABPQR, ACPST	
21	8	8	A, B, C, D	P, Q, R, S, T	ABCD, ABPQRS, ACPQTU	
22	8	16	A, B, C, D	P, Q, R, S, T, U	ABCD, ABPQRS, ACPQTU	
23a	8	16	A, B, C, D	P, Q, R, S, T, U, V	ABCD, ABPQR, ACPST, ADQUV	
23b	8	16	A, B, C, D	P, Q, R, S, T, U, V	ABCD, PQRSTUV, ABPQR, ACRST	

279

Table 16A.3 (continued)

Plan No.	No. of Whole Plots	No. of Split Plots	Whole Plot Factors	Split Plot Factors	Identity Relationship Generators	Confounding* Generator
24	8	16	A, B, C, D	P, Q, R, S, T, U, V, W	ABCD, PQRSTUVW, APQRS, BPQTU, CPRUV	
25	16	4	A, B, C	P, Q, R	ABCRS	PQR
26	16	4	A, B, C	P, Q, R, S	ABPQR, ACPST	PQR
27	16	4	A, B, C	P, Q, R, S, T	ABPQRS, ACPQTU	QT
28	16	8	A, B, C	P, Q, R, S, T, U	ABPQRS, ACPQTU, ABCQRTV	PRT
29	16	8	A, B, C	P, Q, R, S, T, U, V	PQRSTUVW, APQRS, BPQTU, CPRUV	RSV
30	16	8	A, B, C	P, Q, R	ABCDPQR	PQ
31	16	4	A, B, C, D	P, Q, R, S	ABPQR, CDPQS	
32	16	8	A, B, C, D	P, Q, R, S, T	ABCPQR, ABDPST	
33	16	8	A, B, C, D	P, Q, R, S, T, U	APQRS, BPQTU, CDPRT	
34	16	8	A, B, C, D	P, Q, R, S, T, U, V	ABCPQR, BCPST, BDQSU, DPRTV	
35a	16	8	A, B, C, D	P, Q, R, S, T, U, V	PQRSTUV, ABPQR, ACRST, ADPTV	
35b	16	8	A, B, C, D	P, Q, R, S, T, U, V, W	PQRSTUVW, ABPQRS, ACPQTU, ADPTV	
36	16	16	A, B, C, D	P, Q, R, S, T, U, V, W	PQRSTUVW, ABPQRS, ACPQTU, ADPRUV	

* Generator of interactions confounded with whole plot replicates.

8) two-level factors, where $n_1 + n_2 = n (= 5, 6, \ldots, 12)$. In some of the plans the whole plot treatment combinations are replicated, and in some they are not. The plans in which the whole plot treatment combinations are not replicated are obtained by confounding some of the higher order interactions among the split plot factors with some or all of the degrees of freedom associated with whole plot treatments. This type of confounding results in a fractional replicate of the 2^n arrangement. For many of these fractional replicate plans, the number of whole plot treatments is inadequate to permit the estimation of all main effects and two-factor interaction effects of the whole plot factors and also a valid whole plot experimental error. However, even when no estimate of whole plot error is available, knowledge of the relative sizes of whole plot effects is important. All of these plans permit the evaluation of all split plot main effects and two-factor interaction effects, as well as the interactions of the whole plot factors with the split factors, when the three-factor and higher order interactions are negligible. Some of these plans also permit the evaluation of some of the three-factor interactions when the remaining three-factor and higher order interactions are negligible. The plans in which the whole plot treatments are replicated are obtained by confounding some of the higher order interactions among the split plot factors with whole plot replicates and, in some cases, also with some or all of the whole plot treatment effects. The plans in which a split plot interaction effect is confounded only with whole plot replicates are full replicates. The plans in which some interactions among split plot factors are confounded with whole plot replicates and some with whole plot treatment effects are fractional replicates. The plans which contain whole plot replicates permit the evaluation of all whole plot main effects and two-factor interactions, all split plot main effects and two-factor interactions, and all two-factor interactions between one whole plot factor and one split plot factor. The plans vary in the degree to which they permit an evaluation of three-factor interactions.

CHAPTER 17

Conclusions and Apologies

17.1. THE DESIGN AND ANALYSIS OF INDUSTRIAL EXPERIMENTS

I have dealt only with confirmatory experiments, that is to say, with multi-factor trials on operating systems. I have outlined, sometimes at tedious length and with embarrassing simplicity, the steps that must be taken in planning such experiments. You must:

1. Analyze all the relevant data that you have, including literature sources, by some variant of "multiple regression." (There is at least one book on the analysis of miscellaneous, unbalanced, historical data.) This will sometimes (perhaps one time in four) be of direct aid in planning further experimental work.

2. List all the factors you want to vary, with the range of variation and the number of levels that you desire for each.

3. List all the responses you plan to measure, with whatever is known about the precision of measurement of each.

4. Construct an "influence matrix" with each factor of item 2 for a row, and each response of item 3 for a column. Enter in each cell what you know of the effect of that factor on that response.

5. Decide—or guess—which factors may not operate additively, and record these potential two-factor interactions.

6. Decide how many trials (each under different experimental conditions) you can afford to make.

7. See whether the number from item 6 exceeds the number of parameters implied by your estimates in items 2 and 5. If it does not, you will not even be able to study your factors at two levels each.

283

8. Decide whether you will learn anything from a set of trials that includes most factors at two levels, with some at three. If you are willing to consider such a coarse survey of the system's operation, then look for a 2^n3^m subset of the full 2^n3^m factorial which will give you the estimate you want, *with* an excess number of degrees of freedom for studying the data. This excess should be greater than 8, I think, and may be 20 or 60.

9. Having found a plan, inspect it carefully for trials that may be impractical or unworkable. If there are several of these, you may have to shrink some factor ranges, or even subdivide the work into two or more subplans.

10. If the plan is inevitably nested, or/and if some factor is at random levels (being a sample of some population), construct a "dummy analysis of variance table" for your chosen design, to make sure that you have adequate degrees of freedom for judging each effect.

11. Build in all that you know and can afford to study. Randomize over what you do not know, insofar as possible.

12. If raw materials must be accumulated, be sure to have an excess for repeated runs, for following promising leads, and for clearing up confusion from aliasing.

13. Do the same for time and manpower as for raw materials. You are in desperate straits if all your funds, time, and raw materials must go into *one* plan, with final answers required at the end. Reserve, then, one third to one half of your capacities for follow-up.

14. Repeat trials that you suspect are atypical.

These suggestions are not meant as substitutes for any of the *technical* thought and preparation that all careful experimenters take for granted. They are meant to replace old-fashioned notions about one-factor-at-a-time experimentation because of the gains in generality and in precision that are possible.

The weary reader may well feel that this little book contains too much postmortem analysis and not enough material on planning ahead. The balance struck here is necessarily a reflection of my own predilections and limitations. A few novelties have been turned up in experiment design, but most of my own life in statistics has indeed been spent "between engagements," that is, in studying the results obtained by experimenters, and only then trying to help them guess what should be done next.

The broad advice on the analysis of experimental data can be put into a few words: "Verify and revise if necessary the assumptions behind the standard analysis you would like to make." These assumptions are of two kinds. Some concern the "model," whether descriptive or mechanistic, that is, the representing equation with its unknown, presumably fixed, constants; the rest concern the distribution of the random fluctuations, including the uncontrollable appearance of occasional bad values. A number of examples of each sort have been given.

The *model* assumptions most commonly violated (and so indicated by the data) are these:

1. The response function is of the same form in X_1 at all settings of $X_2 \ldots X_k$.
2. The response pattern is the same in all blocks and in all replicates except perhaps for constant offsets.

The distributional assumptions that most frequently fail are:

1. There are no bad values in the data (decimal point errors, or wild values that are unlikely ever to appear again under the same experimental conditions).
2. The variance of y is constant at all X.
3. Observed responses are normally distributed at all X.
4. Random disturbances in successive trials are independent or at least uncorrelated.

The reader who has examined earlier chapters will not need to be told again that, whereas homogeneity of variance (like normality and like statistical independence of successive observations) cannot be proved, heterogeneity (like nonnormality and like serial correlation) can be detected in some sets of data.

17.2. OMISSIONS

There are many omissions in these chapters. Those that weigh most heavily on my conscience are:

1. New work on small $2^n 3^m$ plans [Hoke 1974; Margolin 1968, 1972].
2. J. Mandel's extended models for two-way layouts with interactions [Mandel 1969a., b, 1971].
3. My own work on one-at-a-time plans [Daniel 1973].
4. G. F. Watson's work, with its successors, on group screening [Watson 1961; Patel 1962, 1963].
5. Partially hierarchal plans [Brownlee 1965, Chapter 16; Scheffé, 1958, Sections 5.3, 8.3].

6. Balanced and partially balanced incomplete block plans [Clatworthy 1973].
7. Recent work on experiments with mixtures [Scheffé 1963; Gorman and Hinman 1962; McLean and Anderson 1966].
8. Second-order response surface plans [Box et al. 1954, 1955, 1957; Hunter and Box 1965; Draper and Stoneman 1968, Hill and Hunter 1966].
9. J. W. Tukey's work on exploratory data analysis and on robust analysis in general [Reportedly to be published in 1976].
10. Design of experiments for nonlinear models [Herzberg and Cox 1969].
11. Recent work on the power of common significance tests [Wheeler 1974].

There are others that do not weigh so heavily, including some, I am sure, that are not included because of my lack of awareness.

17.3. NOVELTIES

A large part of the book is entirely standard, following Fisher, Yates, Davies, and Cochran and Cox as closely as possible. I append a list for those who want to proceed quickly to my less standard proposals and operations.

1. Interpretation of 2fi (Section 3.4).
2. Randomization and its limitations (Section 3.6).
3. Residuals in a 3^2 (Sections 4.3, 4.4, 4.5, 10.6).
4. One bad value in a 2^3 (Section 5.12).
5. Rounding in Yates's algorithm (Section 6.5).
6. Interpreting effects and interactions (Section 7.2.5).
7. "Logging" (Section 7.3.2).
8. Dependence of residuals on *factor* levels (Section 7.3.4).
9. Structure in 2fi's; separation of interactions from error (Sections 8.2, 8.3, 8.5, 8.6).
10. Estimating the size of an experiment or sequence (Chapter 9).
11. Minimal blocks of 2 (Sections 10.3.2, 10.3.3, 10.4.1).
12. Against Plackett-Burman two-level designs (Section 13.2).
13. Augmenting 2^n3^m main effect plans (Sections 13.3–13.5).
14. Nonnormal error and components of variance (Section 16.2).
15. Fractional replication in split plots (Section 16.4).

17.4. COOKBOOK OR RESEARCH MONOGRAPH?

Earlier drafts of this work have been criticized for their lack of focus. This criticism has aided me in rewriting many chapters, although usually, it now seems to me, the focus has not been sharpened. One tries to see ahead as broadly as possible, although the word *strategy* has not once been used. One tries to carry each precept, hunch, and suspicion as far as possible, down to the last arithmetical detail, But it often happens that the details of the arithmetic uncover something unforeseen. It is standard scientific expository practice (not mine) to conceal these lucky accidents, and to rewrite as if one had proceeded deductively all the way. "First the hypothesis, then the assumptions, then the mathematics; then stop. Leave the arithmetic to the reader or to the lower orders." I believe it is more instructive and more stimulating to show how one finds out.

I have tried to state some tolerably general findings, and to carry them back to some fairly general recommendations. I have tried to carry each through to concrete numerical examples, demeaning though this may appear. This is, then, a cookbook, but only very amateur cooks—and perhaps some philosophers of cooking—will not know that a good cookbook is used better by a good cook.

It must have been clear long before Section 17.2 that this is not a general handbook, not an exhaustive treatise, and surely not an introductory textbook. But, as Section 17.3 rather immodestly insists, it does contain some new results and so may charitably be called a monograph.

I have tried to avoid sybillic doubletalk. I never write "in a certain sense" or "in essence." I have nowhere issued warnings that such and such must be used very cautiously without an immediate example of what is required for caution. There is, then, I hope, no oracular pretension. The serious reply to the question of the section title is, "A bit of each."

Biobliography and Index of Authors

The boldface numbers at the end of each entry are page numbers in this work.

Addelman, S. and O. Kempthorne (1961), *Orthogonal Main-Effect Plans*, ASTIA Arlington Hall Station, Arlington, Va. **220, 224, 226**

—— (1962a), Orthogonal main effect plans for asymmetrical factorial experiments, *Technometrics*, **4**, No. 1 (February), 21–46. **226**

—— (1962b), Symmetrical and asymmetrical fractional factorial experiments, *Technometrics*, **4**, No. 1 (February), 47–58. **226, 233**

—— (1964), Some two-level factorial plans with split plot confounding, *Technometrics*, **6**, No. 3, 253–258. **272**

—— (1969), Sequences of two-level fractional factorial plans, *Technometrics*, **11**, No. 3, 477–509. **248**

Anderson, R. L. and T. A. Bancroft (1952), *Statistical Theory in Research*, Wiley. **155**

Anscombe, F. J. and J. W. Tukey (1963), The examination and analysis of residuals, *Technometrics*, **5**, No. 2 (May), 141–160. **74**

Blom, G. (1958), *Statistical Estimates and Transformed Beta Variables*, Wiley. **156**

Box, G. E. P. and K. B. Wilson (1951), On the experimental attainment of optimal conditions, *J. Roy. Stat. Soc.*, Series B, **13**, 1–45. **34, 225**

—— (1954), Exploration and exploitation of response surfaces, I, *Biometrics*, **10**, 16–60. **34, 35**

—— and P. V. Youle (1955), Exploration and exploitation of response surfaces, II, *Biometrics*, **11**, 287–322. **35**

—— and J. S. Hunter (1957), Multifactor experimental designs, *Ann. Math. Stat.*, **28**, 195. **35**

—— and J. S. Hunter (1961), The 2^{k-p} fractional factorial designs, I, *Technometrics*, **3**, No. 3, 311–351; II, *Technometrics*, **3**, No. 4, 449–458. **200, 218, 219**

Brownlee, K. A., B. K. Kelly, and P. K. Loraine (1948), Fractional replication arrangements, *Biometrika*, **35** (December), 268–282. **200, 219**

—— (1965), *Statistical Theory and Methodology*, 2nd ed., Wiley. **17, 31, 258, 266, 285**

Chew, V., Ed. (1958), *Experimental Designs in Industry*, Wiley. **35, 221**

Clatworthy, W. H. (1973), *Tables of Two Associate Class Partially Balanced Designs*, National Bureau of Standards, Applied Mathematics Series, No. 63. **171, 182, 195, 286**

Cochran, W. G. and G. M. Cox (1957), *Experimental Designs*, 2nd ed., Wiley. **26, 35, 38, 48, 49, 56, 193, 268**

Connor, W. S. and M. Zelen (1959), *Fractional Factorial Experiment Designs for Factors at Three Levels*, National Bureau of Standards, Applied Mathematics Series, No. 54. **220**

Cox, D. R. (1951, 1952), Some systematic experimental designs, *Biometrika*, **38**, 312. Also, Some recent work on systematic experimental designs, *J. Roy. Stat. Soc.*, Series B, **14**, 211. **251**

—— (1958), *Planning of Experiments*, Wiley, pages 279–281. **251, 268**

Cramér, H. (1946), *Mathematical Methods of Statistics*, Princeton University Press. **6**

Curnow, R. N. (1965), A note on G. S. Watson's paper on group screening, *Technometrics*, **7**, No. 3 (August), 444. **177**

Daniel, C. and E. W. Riblett (1954), A multifactor experiment, *Ind. Eng. Chem.*, **46**, 1465–1468. **217**

—— (1959), Use of half-normal plots in interpreting factorial two-level experiments, *Technometrics*, **1**, No. 4 (November). **149**

—— (1961), Locating outliers in factorial experiments, *Technometrics*, **2**, No. 1, 149–156. **40,**

—— (1962). Sequences of fractional replicates in the 2^{p-q} series, *J. Am. Stat. Soc.*, **58**, 403–429. **7, 246**

—— (1963), Corrigenda to "Sequences . . . ," *J. Am. Stat. Soc.*, **59** (December). **246**

—— and F. Wilcoxon (1966), Factorial 2^{p-q} plans robust against linear and quadratic trends, *Technometrics*, **8**, No. 2, 259–278. **253, 256**

—— and F. S. Wood (1971), *Fitting Equations to Data*, Wiley. **178,**

—— (1973), One-at-a-time plans, *J. Am. Stat. Ass.*, **68**, No. 342, 353–360. **25, 285**

Davies, O. L. and W. A. Hay (1950), Construction and uses of fractional factorial designs in industrial research, *Biometrics*, **6**, No. 3, 233–249. **7, 242**

——, Ed. (1971). *The Design and Analysis of Industrial Experiments*, Hafner (Macmillan). **38, 67, 128, 135, 153, 172, 182, 187, 193, 200, 208, 214, 219, 266**

Draper, N. R. and D. M. Stoneman (1968), Response surface designs for factors at two and three levels, *Technometrics*, **10**, No. 1 (February), 177–192. **286**

Federer, W. T. (1955), *Experimental Design*, Macmillan. **26, 268**

Finney, D. J. (1945), Fractional replication of factorial arrangements, *Ann. Eugen.*, **12**, 291–301. **200, 219**

Fisher, R. A. (1953). *Design of Experiments*, 6th ed., Hafner (Macmillan). **26, 159, 196**

Gaylor, D. W. and F. N. Hopper (1969), Estimating degrees of freedom by Satterthwaite's formula, *Technometrics*, **11**, No. 4, 691–697. **266**

Gnanadesikan, R. and M. B. Wilk (1969), *Multivariate Analysis*, Vol. II (P. R. Krishnaiàh, Ed.), Academic Press, pages 593–638. **218**

Gorman, J. W. and J. E. Hinman (1962), Simplex lattice designs, *Technometrics*, **4**, No. 4, 463–487. **286**

Guttman, I., S. S. Wilks, and J. S. Hunter (1971), *Introductory Engineering Statistics*, 2nd ed., Wiley. **35,**

Hahn, G. J. and S. S. Shapiro (1966), *Catalog and Computer Program for the Design and Analysis of Orthogonal Symmetric and Asymmetric Fractional Factorial Experiments*, Report No. 66-C-165, General Electric Research and Development Center, Schenectady, N. Y., May. **218, 219, 220, 231**

Hald, A. (1952a), *Statistical Tables*, Wiley. **266**

—— (1952b), *Statistical Theory with Engineering Applications*, Wiley, page 510. **251**

Harter, H. (1969), *Order Statistics and Their Use*, Vol. 2, Aeronautical Research Laboratories, page 456. **156**

Herzberg, A. M., and D. R. Cox (1969), Recent work on the design of experiments, *J. Roy. Stat. Soc.*, Series A, **132** pt. 1, 29. **286**

Hicks, C. R. (1965). *Fundamental Concepts in the Design of Experiments*, Holt, Rhinehart, and Winston. **268**

Hill, W. J. and W. G. Hunter (1966). A review of response surface methodology: a literature survey, *Technometrics*, **8**, No. 4 (November), 571–590. **242, 286**

Hoke, A. T. (1970), Research and Development Center, Armstrong Cork Company, Lancaster, Pa. **231**

——— (1974), Economical 3^n plans, *Technometrics*, **16**, No. 3, 375–384. **224, 285**

Hotelling, H. (1944), Some problems in weighing and other experimental techniques, *Am. Math. Stat.*, **15**, 297–306. **4**

Hunter, W. G. and G. E. P. Box (1965), The experimental study of physical mechanisms, *Technometrics*, **7**, No. 1, 23–42. **286**

John, P. W. M. (1973), Personal communication. **65**

Johnson, N. L. and F. C. Leone (1964), *Statistics and Experimental Design*, Vol. 2, Wiley, page 184. **211, 268**

Kempthorne, O. (1952), *Design and Analysis of Experiments*, Wiley. **128, 140, 196**

——— and R. G. Tischer (1953), An example of the use of fractional replication, *Biometrics*, **9**, No. 3, 302–303. **272**

Mandel, J. (1964), *The Statistical Analysis of Experimental Data*, Wiley, Chapter 11. **152**

——— (1969a), A method of fitting empirical surfaces, *Technometrics*, **11**, No. 3 (August). **152, 285**

——— (1969b), The partitioning of interaction in analysis of variance, *J. Res. Natl. Bur. Stan.*, Section B: Mathematical Sciences, **73B**, No. 4. **152, 285**

——— (1971), A new analysis of variance model for non-additive data, *Technometrics*, **13**, No. 1 (February), 1–18. **152**

Margolin, B. H. (1967), Systematic methods for analysing $2^n 3^m$ factorial experiments, *Technometrics*, **9**, No. 2 (May), 245–260. **38**

——— (1968), Orthogonal main-effect $2^n 3^m$ designs and two-factor interaction aliasing, *Technometrics*, **10**, No. 3, 559–573. **229, 231, 285**

——— (1969), Results on factorial designs of Resolution IV for the 2^n and $2^n 3^m$ series, *Technometrics*, **11**, No. 3, 431–444. **231**

——— (1972), Non-orthogonal main-effect designs, *J. Roy. Stat. Soc.*, Series B, **34**, No. 3, 431–440. **285**

McGuire, J. B., E. R. Spangler, and L. Wong (1961), The Size of the Solar System, *Sc. Am.* (April), 64. **23**

McLean, R. A. and V. L. Anderson (1966), Extreme vertices design of mixture experiments, *Technometrics*, **8**, No. 3, 447–454. **286**

Moses, L. E. and R. V. Oakford (1963), *Tables of Random Permutations*, Stanford University Press. **66**

National Bureau of Standards (1950), *Table of Binomial Probability Distribution*, page 211. **137**

National Bureau of Standards (1962), *Fractional Factorial Experiment Designs for Factors at Two Levels*, Applied Mathematics Series, No. 48. **218**

Owen, D. B. (1962), *Handbook of Statistical Tables*, Addison-Wesley, pages 151 ff. **73**

Patel, M. S. (1962, 1963), Group screening with more than two stages, *Technometrics*, **4**, No. 2 (May), 209–217. **177, 285**. Also, A note on Watson's paper, *Technometrics*, **5**, No. 3 (August), 397–398.

Pearson, E. S. and H. O. Hartley (1954), *Biometrika Tables for Statisticians*, Vol. 1. **48**

Pfister Chemical Company (1955), *Naphthols*, Ridgefield, N. J. **162**

Plackett, R. L. and J. P. Burman (1946), Design of optimal multifactorial experiments, *Biometrika*, **23**, 305–325. **224, 231**

Quenouille, M. H. (1953), *The Design and Analysis of Experiment*, Hafner (Macmillan). **193**

Rand Corporation (1955), *A Million Random Digits . . .*, Free Press, Glencoe, Ill. **77**

Satterthwaite, F. E. (1946), An approximate distribution of estimates of variance components, *Biom. Bull.*, **2**, 110–114. **266**

Scheffé, H. (1954), General theory of evaluation of several sets of constants and several sources of variability, *Chem. Eng. Prog.*, **50**, No. 4, 200–205. **275**

⸺ (1958, 1963), Experiments with mixtures, *J. Roy. Stat. Soc.*, Series B, **20**, 344–360. Also, Simplex centroid design for experiments with mixtures, *J. Roy. Stat. Soc.*, Series B, **25**, 235–263. **286**

⸺ (1959), *The Analysis of Variance*, Wiley. **31, 155**

Snedecor, G. W. (1965), *Statistical Methods*, Iowa State University Press, Section 11.14, pages 321–326. **45**

Stefansky, W. (1972a), Rejecting outliers in factorial designs, *Technometrics*, **14**, No. 4, 469–479. **39, 40, 52**

⸺ (1972b), Rejecting outliers by maximum normed residual, *Ann. Math. Stat.*, **42**, No. 1, 34–45. **40**

Tukey, J. W. (1949), One degree of freedom for non-additivity, *Biometrics*, **5**, 232–242. **45**

⸺ (1962). The future of data analysis, *Ann. Math. Stat.*, **33**, 1–67. **156**

Vaurio, V. W. and C. Daniel (1954), Evaluation of several sets of constants and several sources of variability, *Chem. Eng. Prog.*, **50**, No. 3, 81–86. **275**

Watson, G. S. (1961), A study of the group screening design, *Technometrics*, **3**, No. 3 (August), 371–388. **177, 285**

Webb, S. (1965a), *Design, Testing and Estimation in Complex Experimentation*, Vol. I, Office of Aerospace Research, ARL 65–116, Part I, June, page 152. **177**

⸺ and S. W. Galley (1965b) *A Computer Routine for Evaluating Incomplete Factorial Designs*, Office of Aerospace Research, ARL 65–116, Part IV, June. Order No. ED 618518. **231**

⸺ (1971) Small incomplete factorial designs for two- and three-level factors. *Technometrics* **13**, 243–256.

⸺ (1973) Corrigenda to above. *Technometrics* **15**, 951

Wheeler, R. E. (1974–1975), Portable power, *Technometrics*, **16**, No. 2 (May), 193–201. Also, *Technometrics*, **17**, No. 2 (May), 177–179. **286**

Wilk, M. B., and O. Kempthorne (1955), Fixed, mixed, and random models in the analysis of variance, *J. Am. Stat. Ass.*, **50**, 1144–1167. **275**

Yates, F. (1937), *The Design and Analysis of Factorial Experiments*, Bulletin 35, Imperial Bureau of Soil Science, Harpenden, Herts, England, Hafner (Macmillan). **21, 43, 67, 187, 192, 206**

⸺ (1970), *Selected Papers*, Hafner (Macmillan), page 55. **152**

Youden, W. J. (1961), Partial confounding in fractional replication, *Technometrics*, **3**, No. 3 (August), 353–358. **193**

⸺ (1962), *Experimentation and Measurement*, Vistas of Science, National Science Teachers Association, Scholastic Book Series, page 94. **23**

Index